Margaret Plues

Rambles in Search of Wild Flowers, and How to Distinguish Them

Second Edition

Margaret Plues

Rambles in Search of Wild Flowers, and How to Distinguish Them
Second Edition

ISBN/EAN: 9783337034917

Printed in Europe, USA, Canada, Australia, Japan

Cover: Foto ©berggeist007 / pixelio.de

More available books at **www.hansebooks.com**

RAMBLES IN SEARCH

OF

WILD FLOWERS.

RAMBLES IN SEARCH

OF

WILD FLOWERS,

AND

HOW TO DISTINGUISH THEM.

BY

MARGARET PLUES,

AUTHOR OF "RAMBLES IN SEARCH OF FERNS;" "RAMBLES IN SEARCH OF MOSSES,"
ETC., ETC., ETC.

"And the earth brought forth grass and herb yielding seed after his kind, and the tree yielding fruit, whose seed was in itself, after his kind, and God saw that it was good."—GEN. I. 12.

SECOND EDITION.

LONDON:
JOURNAL OF HORTICULTURE AND COTTAGE GARDENER OFFICE,
171, FLEET STREET, E.C.
1864.

LONDON:
PRINTED AT THE HORTICULTURAL PRESS,
171, FLEET STREET.

DEDICATION.

In memory of the pleasant conversations on Wild Flowers, which ever followed upon our returning blossom-laden from childish rambles through field and wood, and which sowed in our minds a love for the study of Nature, this book is dedicated with tender reverence and gratitude

TO MY MOTHER.

> " Thus do we think of her, and keep unbroken
>   The bond that Nature gives;
> Thinking that our remembrance, though unspoken,
>   May reach her where she lives."

TABLE OF CONTENTS.

CLASS I.—DICOTYLEDONS.

Subclass I.—THALAMIFLORALS.

1. Ranunculaceæ . . . 11
   Genera.—1, Clematis. 2, Thalictrum. 3, Anemone. 4, Adonis. 5, Ranunculus. 6, Myosurus. 7, Trollius. 8, Caltha. 9, Helleborus. 10, Actæa. 11, Aquilegia. 12, Aconitum. 13, Delphinium. 14, Pæonia.
2. Nymphaceæ . . . 22
   Genera.—1, Nymphæa. 2, Nuphar.
3. Papaveraceæ . . . 23
   Genera.—1, Papaver. 2, Meconopsis. 3, Glaucium. 4, Chelidonium majus.
4. Fumariaceæ . . . 25
   Genera.—1, Fumaria.
5. Berberidaceæ . . . 26
   Genera.—1, Berberis. 2, Epimedium.
6. Cruciferæ . . . 28
   Genera. — 1, Hutchinsia. 2, Thlaspi. 3, Capsella. 4, Teesdalia. 5, Lepidium. 6, Cochlearia. 7, Subularia. 8, Draba. 9, Cakile. 10, Crambe. 11, Senebiera. 12, Alyssum. 13, Iberis. 14, Dentaria. 15, Cardamine. 16, Arabis. 17, Turritis. 18, Barbarea. 19, Nasturtium. 20, Sisymbrium. 21, Erysimum. 22, Cheiranthus. 23, Mathiola. 24, Brassica. 25, Sinapis. 26, Raphanus. 27, Hesperis.
7. Resedaceæ . . . 43
   Genera.—1, Reseda.
8. Cistaceæ . . . 44
   Genera.—1, Cistus.
9. Violaceæ . . . 45
   Genera.—1, Viola.
10. Droseraceæ . . . 47
    Genera.—1, Drosera. 2, Parnassia.
11. Polygalaceæ . . . 48
    Genera.—1, Polygala.
12. Tamaricaceæ . . . 49
    Genera.—1, Tamarisk.
13. Frankeniaceæ . . . 49
    Genera.—1, Frankenia.
14. Elatinaceæ . . . 49
    Genera.—1, Elatine.
15. Caryophyllaceæ . . 50
    Genera. — 1, Dianthus. 2, Silene. 3, Saponaria. 4, Lychnis. 5, Agrostemma. 6, Sagina. 7, Mœnchia. 8, Holosteum. 9, Spergula. 10, Stellaria. 11, Arenaria. 12, Cerastium.

	PAGE		PAGE
16. TILIACEÆ	60	20. ACERACEÆ	68

16. TILIACEÆ 60
GENERA.—1, Tilia.

17. HYPERICACEÆ . . . 61
GENERA.—1, Hypericum.

18. MALVACEÆ 63
GENERA.—1, Malva. 2, Althæa. 3, Lavatera.

19. LINACEÆ 66
GENERA.—1, Linum. 2, Radiola.

20. ACERACEÆ 68
GENERA.—1, Acer.

21. GERANIACEÆ. . . . 69
GENERA.—1, Geranium. 2, Erodium.

22. BALSAMINACEÆ . . . 73
GENERA.—1, Impatiens.

23. OXALIDACEÆ. . . . 73
GENERA.—1, Oxalis.

SUBCLASS II.—CALYCIFLORALS.

1. CELASTRACEÆ 78
GENERA.—1, Euonymus.

2. RHAMNACEÆ 79
GENERA.—1, Rhamnus.

3. LEGUMINOSÆ 80
GENERA.—1, Ulex. 2, Genista. 3, Cytisus. 4, Ononis. 5, Medicago. 6, Melilotus. 7, Trifolium. 8, Lotus. 9, Oxytropus. 10, Anthyllis. 11, Astragalus. 12, Vicia. 13, Ervum. 14, Ornithopus. 15, Hippocrepis. 16, Hedysarum. 17, Lathyrus. 18, Pisum. 19, Orobus.

4. ROSACEÆ 103
GENERA.—1, Prunus. 2, Spiræa. 3, Dryas. 4, Geum. 5, Potentilla. 6, Tormentilla. 7, Fragaria. 8, Rubus. 9, Agrimonia. 10, Alchemilla. 11, Sanguisorba. 12, Poterium. 13, Rosa. 14, Pyrus. 15, Mespilus. 16, Cratægus.

5. ŒNOTHERACEÆ . . . 109
GENERA.—1, Isnardia. 2, Circæa. 3, Epilobium. 4, Œnothera.

6. HIPPURIDACEÆ . . . 112
GENERA.—1, Hippuris. 2, Myriophyllum. 3, Callitriche. 4, Ceratophyllum.

7. LYTHRACEÆ 113
GENERA.—1, Lythrum. 2, Peplis.

8. CUCURBITACEÆ . . . 114
GENERA.—1, Bryonia.

9. PORTULACEÆ . . . 115
GENERA.—1, Montia.

10. ILLECEBRACEÆ . . 118
GENERA.—1, Herniaria. 2, Illecebrum. 3, Polycarpon. 4, Scleranthus.

11. CRASSULACEÆ . . . 120
GENERA.—1, Tillæa. 2, Cotyledon. 3, Sempervivum. 4, Sedum. 5, Rhodiola.

12. GROSSULARIACEÆ . . 123
GENERA.—1, Ribes.

13. SAXIFRAGACEÆ . . 124
GENERA.—1, Saxifraga. 2, Chrysosplenium.

14. UMBELLIFERÆ . . . 129
GENERA.—1, Scandix. 2, Anthriscus. 3, Chærophyllum. 4, Eryngium. 5, Sanicula. 6, Daucus. 7, Torilis. 8, Myrrhis. 9, Bunium. 10, Œnanthe. 11, Crithmum. 12, Athanotis. 13, Pimpinella.

14, Sium. 15, Sison. 16, Cicuta. 17, Conium. 18, Angelica. 19, Smyrnium. 20, Apium. 21, Meum. 22, Ægopodium. 23, Carum. 24, Cnidium. 25, Bupleurum. 26, Hydrocotyle. 27, Æthusa. 28, Imperatoria. 29, Selinum. 30, Ligusticum. 31, Pastinaca. 32, Peucedanum. 33, Tordylium. 34, Heracleum.

15. ARALIACEÆ 142
GENERA.—1, Hedera. 2, Adoxa.

16. CORNACEÆ 143
GENERA.—1, Cornus.

SUBCLASS III.—COROLLIFLORALS.

1. VISCACEÆ 146
GENERA.—1, Viscum.

2. CAPRIFOLIACEÆ . . . 148
GENERA.—1, Sambucus. 2, Viburnum. 3, Lonicera. 4, Linnæa.

3. RUBIACEÆ 150
GENERA.—1, Rubia. 2, Galium. 3, Asperula. 4, Sherardia.

4. VALERIANACEÆ . . . 154
GENERA.—1, Valeriana. 2, Fedia.

5. DIPSACACEÆ . . . 155
GENERA.—1, Dipsacus. 2, Scabiosa.

6. COMPOSITÆ 158
GENERA.—1, Tragopogon. 2, Picris. 3, Apargia. 4, Thrincia. 5, Hypochœris. 6, Hieracium. 7, Lactuca. 8, Sonchus. 9, Crepis. 10, Leontodon. 11, Lapsana. 12, Cichorium. 13, Arctium. 14, Serratula. 15, Carduus. 16, Cnicus. 17, Onopordum. 18, Carlina. 19, Centaurea. 20, Bidens. 21, Chrysocoma. 22, Eupatorium. 23, Diotis. 24, Tanacetum. 25, Artemisia. 26, Gnaphalium. 27, Petasites. 28, Tussilago. 29, Erigeron. 30, Solidago. 31, Senecio. 32, Doronicum. 33, Conyza. 34, Inula. 35, Bellis. 36, Chrysanthemum. 37, Pyrethrum. 38, Matricaria. 39, Anthemis.

7. CAMPANULACEÆ . . 180
GENERA.—1, Campanula. 2, Phyteuma. 3, Jasione.

8. LOBELIACEÆ 184
GENERA.—1, Lobelia.

9. VACCINIACEÆ . . . 185
GENERA.—1, Vaccinium.

10. ERICACEÆ 185
GENERA.—1, Erica. 2, Calluna. 3, Menziesia. 4, Azalea. 5, Andromeda. 6, Arbutus.

11. MONOTROPEÆ . . . 189
GENERA.—1, Pyrola. 2, Monotropa.

12. AQUIFOLIACEÆ. . . 189
GENERA.—1, Ilex.

13. OLEACEÆ 190
GENERA.—1, Ligustrum. 2, Fraxinus.

14. APOCYNACEÆ . . . 194
GENERA.—1, Vinca.

15. GENTIANEÆ . . . 195
GENERA.—1, Gentiana. 2, Erythræa. 3, Exacum. 4, Chlora. 5, Menyanthes. 6, Villarsia.

16. POLEMONIACEÆ . . 198
GENERA.—1, Polemonium.

17. CONVOLVULACEÆ . 199
GENERA.—1, Convolvulus. 2, Cuscuta.

18. BORAGINEÆ . . . 200
GENERA.—1, Echium. 2, Pulmonaria. 3, Lithospermum. 4, Symphytum. 5, Borago. 6, Lycopsis. 7, Anchusa. 8, Myosotis. 9, Asperugo. 10, Cynoglossum.

19. SOLANACEÆ. . . . 206
GENERA.— 1, Solanum. 2, Atropa. 3, Hyoscyamus.

20. OROBANCHACEÆ . . 210
GENERA. — 1, Orobanche. 2, Lathræa.

21. SCROPHULARIACEÆ . 211
GENERA.—1, Digitalis. 2, Antirrhinum. 3, Linaria. 4, Scrophularia. 5, Limosella. 6, Melampyrum. 7, Pedicularis. 8, Rhinanthus. 9, Bartsia. 10, Euphrasia. 11, Sibthorpia. 12, Veronica. 13, Verbascum.

22. LABIATÆ 220
GENERA.—1, Lycopus. 2, Salvia. 3, Mentha. 4, Thymus. 5, Origanum. 6, Ajuga. 7, Teucrium. 8, Ballota. 9, Leonurus. 10, Galeobdolon. 11, Galeopsis. 12, Lamium. 13, Betonica. 14, Stachys. 15, Nepeta. 16, Glechoma. 17, Marrubium. 18, Acinos. 19, Calamintha. 20, Clinopodium. 21, Melittis. 22, Prunella. 23, Scutellaria.

23. VERBENACEÆ . . . 229
GENERA.—1, Verbena.

24. LENTIBULARIACEÆ . 230
GENERA.—1, Pinguicula. 2, Utricularia.

25. PRIMULACEÆ . . . 231
GENERA.—1, Primula. 2, Hottonia. 3, Cyclamen. 4, Anagallis. 5, Lysimachia. 6, Centunculus. 7, Trientalis. 8, Glaux. 9, Samolus.

26. PLUMBAGINEÆ . . 238
GENERA.—1, Statice.

27. PLANTAGINACEÆ . . 238
GENERA.—1, Plantago. 2, Littorella.

SUBCLASS IV.—APETALS.

1. CHENOPODIACEÆ . . 240
GENERA.—1, Chenopodium. 2, Atriplex. 3, Beta. 4, Salsola. 5, Salicornia.

2. POLYGONACEÆ . . . 244
GENERA.—1, Polygonum. 2, Rumex. 3, Oxyria.

3. ELEAGNEÆ 247
GENERA —1, Hippophæ.

4. THYMELACEÆ . . . 248
GENERA.—1, Daphne.

5. SANTALACEÆ . . . 248
GENERA.—1, Thesium.

6. ARISTOLOCHIACEÆ . . 249
GENERA.—1, Aristolochia. 2, Asarum.

7. EMPETRACEÆ . . . 249
GENERA.—1, Empetrum.

8. EUPHORBIACEÆ . . 250
GENERA.—1, Euphorbia. 2, Mercurialis. 3, Buxus.

9. URTICACEÆ 252
GENERA.—1, Urtica. 2, Parietaria. 3, Humulus.

10. ULMACEÆ 256
GENERA.—1, Ulmus.

11. SALICACEÆ 257	14. CORYLACEÆ 263
GENERA.—1, Salix. 2, Populus.	GENERA.—1, Fagus. 2, Castanea. 3, Quercus. 4, Corylus. 5, Carpinus.
12. MYRICACEÆ. . . . 262	
GENERA.—1, Myrica.	15. CONIFERÆ 266
13. BETULACEÆ . . . 260	GENERA.—1, Pinus. 2, Juniperus. 3, Taxus.
GENERA.—1, Betula. 2, Alnus.	

CLASS II.—MONOCOTYLEDONS.

SUBCLASS I.—PETALLIDS.

1. HYDROCHARIDACEÆ . 272	8. MELANTHACEÆ . . . 291
GENERA.—1, Hydrocharis. 2, Stratiotes. 3, Anacharis.	GENERA.—1, Colchicum. 2, Tofieldia.
2. ORCHIDACEÆ . . . 274	9. JUNCACEÆ 292
GENERA.—1, Orchis. 2, Gymnadenia. 3, Habenaria. 4, Listera. 5, Neottia. 6, Ophrys. 7, Epipactis. 8, Goodyera. 9, Corallorhiza. 10, Malaxis. 11, Liparis. 12, Cypripedium.	GENERA.—1, Juncus. 2, Luzula. 3, Nartheceum.
	10. BUTOMACEÆ . . . 296
	GENERA.—1, Butomus.
	11. ALISMACEÆ. . . . 296
3. IRIDACEÆ 279	GENERA.—1, Alisma. 2, Sagittaria.
GENERA.—1, Iris. 2, Trichonema. 3, Crocus.	
	12. JUNCAGINACEÆ . . 297
4. AMARYLLIDACEÆ . . 281	GENERA.—1, Triglochin.
GENERA.—1, Narcissus. 2, Galanthus. 3, Leucojum.	13. TYPHACEÆ 298
	GENERA.—1, Typha. 2, Sparganum.
5. DIOSCORIACEÆ . . . 285	14. ARACEÆ 299
GENERA.—1, Tamus.	GENERA.—1, Arum.
6. TRILLIACEÆ 285	15. ORONTIACEÆ . . . 300
GENERA.—1, Paris.	GENERA.—1, Acorus.
7. LILIACEÆ 285	16. PISTIACEÆ 301
GENERA.—1, Asparagus. 2, Ruscus. 3, Convallaria. 4, Hyacinthus. 5, Scilla. 6, Ornithogalum. 7, Allium. 8, Gagea. 9, Tulipa. 10, Fritillaria. 11, Anthericum.	GENERA.—1, Lemna.
	17. NAIADACEÆ . . . 302
	GENERA.—1, Potamogeton. 2, Ruppia. 3, Zannichellia. 4, Zostera.

Subclass II.—GLUMALS.

1. CYPERACEÆ 306

   GENERA.—1, Schœnus. 2, Rhyncospora. 3, Cyperus. 4, Scirpus. 5, Eleocharis. 6, Eriophorum. 7, Carex.

2. GRAMINACEÆ . . . 312

   GENERA.— 1, Anthoxanthum. 2, Nardus. 3, Phalaris. 4, Phleum. 5, Alopecurus. 6, Polypogon. 7, Milium. 8, Agrostis. 9, Cynodon. 10, Digitaria. 11, Panicum. 12, Aira. 13, Holcus. 14, Hierochloe. 15, Melica. 16, Sesleria. 17, Glyceria. 18, Poa. 19, Briza. 20, Dactylis. 21, Spartina. 22, Cynosurus. 23, Festuca. 24, Bromus. 25, Avena. 26, Lolium. 27, Lagurus. 28, Arundo. 29, Ammophila. 30, Rottbollia. 31, Elymus.

RAMBLES

IN SEARCH OF WILD FLOWERS,

AND

HOW TO DISTINGUISH THEM.

INTRODUCTORY.

> " Enshrined within the tiny flowers
> That grow beside the path of life
> Are simples blest with healing powers,
> And germs with sweetest odour rife.
> But he alone that stooping low,
> Will stay with curious hand to cull,
> Can all the many virtues know
> That dignify the Beautiful."
>
> PNEUMA.

THE unusually mild winter of 185— had induced many families to pass it at the sea-side, and the inhabitants of Clevedon, in Somersetshire, numbered more than the regular residents. Mrs. Dring, a widow lady, and her only daughter were occupying a handsome suite of rooms in one of the houses on the west of the town, and their Christmas had passed cheerily, enlivened by the society of two cousins, Esther Claridge, the eldest daughter of Mrs. Dring's brother, and Edward Leigh, the son of her sister.

Two seasons in London, followed by gay autumns at watering-places, had made sad havoc with the health of Fanny Dring, and her pale face and attenuated figure, as she stood in the large bay window gazing wistfully on the lovely landscape, attracted her mother's anxious observation.

"I am very glad we decided to come here," she said, as she gathered up some letters and prepared to leave the room to write the answers; "I am certain that this air suits Fanny. It did not seem to take any effect upon her the first fortnight; but since Christmas she has decidedly improved. Ere summer comes my child will be bright and gay again!"

As her mother closed the door, Fanny's eye still dwelt on the outward view. She was not gazing on the beach with its small complement of pedestrians, nor upon the waters of the Severn, with which the fresh tide of the channel was now striving, nor yet upon the grey line of the Welsh coast and the scarcely less faint outline of the flat and steep Holmes; but her eye was fixed with a yearning look of inquiry upon the distant thread where sky and water meet, and that look seemed to ask, "What is there in the future for me? Is there rest? for my spirit is weary of all that has been."

She turned slowly as her mother's footsteps receded, and addressed Esther.

"My mother is quite mistaken," she said gloomily. "It is not the air of Clevedon that has done me good. True, I am able to walk more, and I have enjoyed our strolls of late; but that is only because you found things constantly to interest me. Every common shell stranded on the coast, the weed in the tide-pools and the slimy things playing hide-and-seek among it, each little belated flower, and every insect crawling on the face of the cliff, formed an object of interest and pleasure seen through your eyes. But now that you are leaving us, my life will become dreary and aimless again, and my daily walk a dull cold task. The good influence of Clevedon will end, as it began, with you."

"I can quite enter into your feelings, Fanny," broke in Edward, throwing aside his book and joining his cousins at the window. The boy was seventeen, three years younger than Fanny, and ten younger than the staid Esther. He was a brilliant scholar in a large public school in one of the midland counties, and was spending his holidays with his aunt.

"My mother is always exhorting me to take exercise," he continued, "and I have no objection to a good game at cricket; but for a constitution walk!—bah!—it is the greatest bore that ever was invented. But whilst I have been here, I have always liked our rambles; the things that Esther finds amuse me, more especially those jolly little beasts which we find in the pools among the rocks. If I lived with her I should be a much more dutiful son, for I would take a walk as often as my mother wished."

While the boy spoke, Esther was weighing Fanny's words. "Your mind is in great need of interesting occupation, Fanny," she said. "How I wish I could make over some of my super-abundant work to you! You would never be in want of objects of interest if you had brothers to care for and young sisters to teach. I can scarcely get time to write letters, or to read for my own information. Some of the time which hangs so heavily on your hands would be a great boon to me!"

"I would help you most willingly if I could," Fanny replied; "but I cannot fly to your Yorkshire home to give music lessons, and back to my mother by evening. I do not love idleness, and I am weary for want of an object for exertion; that would revive me more than all the tonics in the world, or the finest and most frequent changes of air. I have thought of commencing the study of a language; but my spirit sinks at the idea of close solitary labour, and what better should I be when I had succeeded?"

"Let me advise you, Fanny, to take up some natural science instead. There is a mental nourishment to be found in the study of God's works. His glorious creation is outspread, and its testimony is proclaimed to the simplest observer; but the more closely we look into His works the more fully are we penetrated by their beauty, and many a message of Divine love and peace comes to us by their 'silent lips.' You are very fond of flowers—begin to study them. A little patient investigation will enable you to understand their structure, habits, and classification; and if you will diligently arrange and collect, you may be of real and efficient help to me. Last year I undertook a work too great for the small amount of time I have at my disposal. My uncle, the Vicar of ——— in Devonshire, is forming a scientific institution in connection with the Mill schools, which are of a high order. The millowner is a truly benevolent and enlightened man—he does not regard his operatives as mere 'hands,' and he is as anxious as my uncle to promote interest and recreation for them after working hours. Uncle Henry means to give botanical lectures next winter, hoping thereby to provide the poor people with mental occupation at the time, and to induce them to spend their occasional holidays during the succeeding summer in the country; and I have undertaken to make a collection of dried plants for illustration, with gigantic sketches of one member of each natural family; also to gather all collateral information and association which may enliven the subject to the young mechanics. If you and Edward will help me, I shall be most thankful; you could easily make the drawings, and both could collect specimens of any plants you find in flower during your rambles in the fields and copses."

"I will help you, with all my heart," replied Fanny; "but I know absolutely nothing of botany. Will you, during the next few days, try to teach me the necessary principles?"

"Oh! yes, do, rejoined Edward, "and admit me into the

class! I scarcely knew a Dandelion from a Buttercup when I came here, and now I know the Michaelmas Daisy, and the Sea-Orache, and the Knot-grass, and ever so many other things. There are jolly plants by the Avon, and heaps of bothering weeds choking up the stream, and I know a lane in Shropshire where garden flowers grow on the banks. We find some knobby stories about flowers in our school books, which will make the young operatives open their eyes very wide. Oh, I shall be able to help in a first-rate way, if only you can knock the principles of the science into my stupid head!"

Esther took a few minutes to collect and clear her thoughts, and then she began her first lesson, Edward gravely taking notes as in a lecture-room.

"All plants, from the forest tree to the microscopic Fungus, are divided into three classes, according to the form of the seed:—I. The Two-LOBED (Dicotyledonous). II. The ONE-LOBED (Monocotyledonous). III. The LOBELESS (Acotyledonous). The greatest part of our trees and plants belong to the first class—the Two-lobed; bulbous plants, water plants, Grasses, Sedges, and a few other families, belong to the One-lobed or second class; Ferns, Mosses, Lichens, Sea-weeds, and Fungi, belong to the third or Lobeless class. The first two classes contain the FLOWERING PLANTS, the third the FLOWERLESS PLANTS. If you take a Bean and a grain of corn and keep them in a warm, moist place, you will see the Bean open in two valves or *lobes* (*fig.* 1, A), and a small bud will arise from the lower end, which is the embryo of the new plant (*fig.* 1, B). This proves the Bean to be a member of the first, or Two-lobed class. The corn, on the other hand, makes no division—it has only one lobe; roots push out at the lower end; and the bud, containing stem, leaves, and flower, shoots from the upper. It stands for an

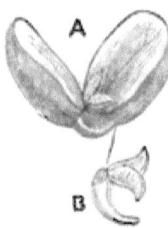

Fig. 1.

example of the One-lobed class. *Fig.* 2 represents a germinating seed of Indian Corn—*a* is the one lobe, perforated by

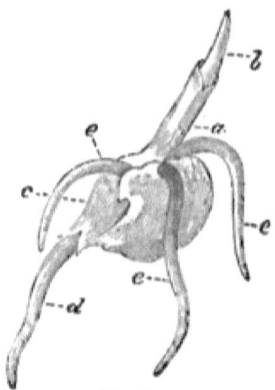

Fig. 2.

the plumule *b*; *c* is the root-sheath; *d* the radicle; and *e e e* the adventitious roots. Supposing the seedling to have become a perfect and mature plant, we find it possessed of six distinct parts—root, stem, leaves, flower, fruit, and seed. The office of the root is to draw nourishment from the ground, while the leaves answer to the breathing organs of animals. The principal divisions of the classes depend upon the form and arrangement of the flower, so before proceeding further we must consider its parts. There are some blossoms of the Lesser Celandine upon the lawn (*fig.* 3). Will you bring me one, Edward, and I will show you its formation?

"These greenish leaves which enfold the flower when in bud are called *sepals*. Collectively they form the *calyx*. The bright yellow glossy leaves forming the largest portion of the flower are *petals* singly, collectively the *corolla*. This tassel inside the flower is formed of an

Fig. 3.

indefinite number of *pistils* and *stamens*, the former being generally called *carpels*; the latter are composed of the *filaments*, and the heads or *anthers*, in which the *pollen* is contained. The carpels, as well as the stamens, are seated on the receptacle, and their summits are called *stigmas*. You cannot very well distinguish the parts of the pistils in that one of the Celandine; but if you examine the little pillar-looking body that is found in the centre of the flower of the Primrose you will find it to resemble a drumstick. The round end or base is called the germ or *ovary*, the stalk of the pillar is the *style*, and the top or capital is the *stigma*. There are generally *nectaries* or honey-cups situated near the base of the petals of the Celandine."

Here the pupils declared that they had heard enough for the first lesson. They would con it over again and make it fully their own, and would ask further information on the morrow.

"Now, Esther," said Fanny, as soon as the breakfast-cloth was withdrawn on the following morning, "be so good as to proceed to the next lesson in botany. We remember the great divisions into Two-lobed, One-lobed, and Lobeless classes, and the six parts of the plant; and the calyx, corolla, stamens, pistil, and germen. What comes next?"

"The next division is into subclasses," began Esther. "These are decided according to the various arrangements of the parts of the flower. In the first subclass the stamens and petals are inserted into the part of the stem beneath the ovary, which is called the *receptacle* or *thalamus*. In this subclass the petals are always distinct; the plants belonging to this group are called THALAMIFLORALS.

"In the second subclass the petals and stamens are inserted on the calyx or close to its base, and the petals are generally distinct. These plants are called CALYCIFLORALS.

"In the third subclass the stamens are fixed either upon the petals, or inserted on the receptacle, as is the case in the first subclass; and the petals are no longer distinct, but joined together so as to form what is called a *monopetalous corolla*. These plants are called COROLLIFLORALS.

"In the fourth subclass either the calyx or corolla, or both, are wanting; the plants are called APETALS. These four subclasses comprise the great class of Dicotyledons, or Two-lobed plants. The two first are many-petalled, the third one-petalled, and the fourth petalless.

"The One-lobed class is but once divided, the first subclass having petals, and the plants being called PETALIDS;

and the second subclass having, in lieu of petals, chaffy scales, or *glumes*, and so being named GLUMALS. These are comprised of Sedges and Grasses."

The young companions were soon called upon to separate; but, before doing so, they agreed to study botany carefully, searching diligently for plants which they should identify if possible, and press carefully. Working thus during the whole spring, summer, and autumn, they should meet the next Christmas at one of their homes, or, still better, at Mr. Carew's, the clergyman for whom the herbarium was to be made, and show all the plants to him, telling him at the same time of all their rambles in search of them, and detailing all the information they had collected. At Edward's suggestion they formed themselves into a botanical club, Esther being President, Fanny Secretary, and Ned Vice-president.

Spring lingered long ere making its appearance that year. Heavy rains during February and high winds in March kept back vegetation; but April, with its smiles and tears, succeeded in alluring the budding plants. The young botanists eagerly availed themselves of the first opportunity for commencing their search, and their efforts did not long remain unrewarded. Fanny might be seen, with an expression of lively interest on her countenance, climbing the cliffs, penetrating the woods, and exploring the salt marshes. Her mother thanked God for the renewed health which tinged her cheek and gave elasticity to her step, and she gladly procured stronger boots and dresses of firm texture in which she might ramble and climb at her own free will without fear of detriment. She made no objection even to the large flower-press, which would have been thought too uncouth for many a less elegant drawing-room; nay, she quite loved the rough machine as a means of procuring health and interest to her daughter.

Matters did not go quite as smoothly with Edward Leigh; but the boy was largely endowed with firmness and held steadily to his purpose. When first the other day-pupils found that he was addicting himself to collecting flowers, they named him "Miss Flora." He declined ever replying to this cognomen, but did not otherwise notice it. A reserved timid boy of the name of Luscombe, who had long been a humble admirer of Edward's, asked to be allowed to join him in collecting, and, after awhile, one of the tutors began to be a member of their Saturday excursions. The three were caricatured upon the great school-gates, and Luscombe was frightened out of his allegiance for some weeks; but the petty persecution died a natural death, and then the rambles regained their charm, even to the timid boy.

And Esther, with her young group of pupil sisters, collected almost daily. Alice and Kate loved to vie with each other as to how many different flowers they could collect in each day's ramble; and the quiet Herbert found his chief amusement in observing the different parts of plants and their curious structure; and he justified his admission to the club by presenting good selections of leaves, stems, roots, and corolla, and making careful sections of the stems of the two great classes. As the summer progressed, Esther had the pleasure of a trip to Scotland, which, with two or three little visits, gave her longer holidays than usual. Fanny and her mother sojourned in Wilts, Devon, and Cornwall, after leaving Clevedon; and Edward divided his summer holidays between friends in Kent and Shropshire.

The flowers disappeared, and fruits came in their stead; the song of the harvest-home sounded as cheerily as the laugh of boys rifling the copse of its nuts. An early snow-storm swept over the cleared fields and put a stop to everything; and then the old year died away, having waited like holy Simeon to behold the birth of the Lord's Christ, and to hear the song of the angels, "Glory to God in the highest, on earth peace, good will to men."

When each member of the botanical club had done honour

to the plum-pudding in their respective homes, they commenced their journey to Mr. Carew's house in Devonshire, Esther pausing in her route to join her cousin Edward, that they might proceed together.

Hearty was the greeting and eager the delight of the young people when each produced his or her packet of dried flowers, and prepared to arrange them together in natural groups. Mr. Carew's daughters, and some young friends staying with them, begged to be allowed to be present, while a clever lad from the school promised to take down all the information in shorthand for the use of the promised lectures. At Esther's desire, each arranged the members of the first natural family, that of the Ranunculus, and so the budgets were fairly opened.

CHAPTER I.

RANUNCULÁCEÆ.

> "Exhibit on thy dress whene'er they bloom
> The Buttercup and Daisy. They will be
> The types of heaven, and holier than the plumes
> A hero wears, and they will preach to thee
> Of how the sun and showers drop favours ceaselessly.
>
> . . . .
>
> "Away! The counterfeit shake off from thee;
> And Nature's gold and silver gather up,
> Such as keeps innocent the child and bee,
> And which from heaven the angels have let drop—
> The Daisy, and her sister flower the Buttercup."
>
> E. H. BURRINGTON.

I SUPPOSE we are all bearing in mind, began the Lady President, that the distinctive feature of the first or Receptacle subclass, or that of the Thalamiflorals, is that the various parts of the flower—calyx, corolla, and stamens, are fastened on the receptacle beneath the germen. The first order in this subclass is that of the RANUNCULÁCEÆ, of which the Buttercup is the most universally-known representative. The general characteristics of this numerous order are from three to six petals and sepals, numerous stamens, and several distinct carpels. They are generally herbs with divided leaves, the footstalk of which spreads out so as to form a sheath, which clasps the stem. Most of them are in a great degree poisonous if taken internally, and painful if applied to the skin. Their numerous stamens and cupped corolla give them a close resemblance to the wholesome Rose order; but the poisonous Ranunculáceæ may at once be distinguished by the early falling of their calyx-leaves, while those of the Rosaceæ continue to adorn the ripening fruit.

The wild Clématis, or Traveller's Joy, represents the first family in the Ranúnculus order. It is an elegant climbing shrub, bearing numerous flowers with a calyx of four parchment-coloured sepals, no corolla, numerous stamens and carpels, which latter become adorned with a long feathery plume, so that the seed-tufts form an even more elegant object than the flowers. Such, at least, was Bishop Mant's opinion—

> "The Traveller's Joy,
> Most beauteous when its flowers assume
> Their autumn form of feathery plume."

I have gathered this plant near Warminster in Wilts, and Fanny has specimens from Devonshire; I have also observed it in travelling along the South-Eastern Railway. But it was in the lanes about Ross in Herefordshire that I found it in the greatest beauty, throwing its frail branches from the supporting arms of trees on the one side, to the tall hedge on the other, and converting the deep lane into a cloistered aisle rich in floral tracery. Its leaves are composed of five leaflets. The flowers appear in July, and the whole shrub is poisonous. (*Plate I., fig.* 1).

The Meadow Rue (Thalíctrum), the next family in the order, is also without corolla. Its four or five sepals are very insignificant, but its clusters of yellow anthers, disposed freely along its branched flower-stalk, have a light appearance, and contrast agreeably with its twice-divided (*bi-pinnate*) dark green leaves. I found the Greater Meadow Rue last July at Mackershaw, a part of the grounds adjoining those of Studley, upon the property of Lord De Grey and Ripon. It grew upon a shady bank, at the foot of which a pretty stream flows.

There are an alpine Meadow Rue, a common Meadow Rue, and a lesser species, all occasionally found in more or less alpine meadows; but we have none of us succeeded in procuring these.

We have all specimens of the Anemóne, or Wind-Flower. This represents the third family of the order. Here, again, the sepals are white or coloured so as to fill the place of the absent corolla, and the carpels become feathered so that the seed has a cottony appearance.

In olden times the Wood Anemone (A. nemorósa, *Plate I.*, *fig.* 2), was supposed to have wonderfully healing properties. The blossom was plucked as soon as it opened. At the same time these words were pronounced, "I gather thee as a remedy against all disease." It was then tied round the neck of the invalid. This pretty plant is familiar to every one who cares to notice wild flowers. In April and May it covers the woods in every sheltered part of England with fleecy whiteness, nodding to every breeze, and gladdening the heart of all who have longed for spring. In more exposed situations—mountainous pastures, alpine woods, &c., it comes later and lasts longer; the flowers are smaller, the divided leaves darker, and the outside of the sepals often tinged with purple.

The Blue Anemone, or Wind-Flower (A. apennína), is found only in Wales, and that very rarely; but it is a common garden flower. The sepals are much more numerous and of a narrower form than in any other species, and their delicate blue colour makes it deservedly a favourite.

The Pasque-Flower Anemone (A. pulsatilla, *Plate I., fig.* 3), is a handsome and well-marked species, growing on chalky uplands. The purple sepals are large and hairy, and form a handsome, upright, bell-shaped flower; the leaves, too, are hairy, and the seeds are adorned with long feathery tails resembling those of the Traveller's Joy.

There is a Yellow Anemone which favours woods in Herefordshire. Its form resembles that of the White one; but its full colour and glossy texture indicate a relationship to the Buttercup. I gathered it abundantly in woods on the banks of the Rhine some years ago, and this season it was found in the richly wooded district of Rolvenden, in Kent. Its foliage is like that of the Wood Anemone.

The beautiful Scarlet Anemone, the superb ornament of

Italian woods, and also of our gardens, is a much-valued member of this family.* It is told of one of the Mayors of our large towns that when first this flower was introduced into our English gardens he exceedingly coveted some of the seed. This he offered to buy, but the monopolist refused to sell: so his Worship tried *finesse* to gain his purpose, and proposed to the florist that, as he would not part with his Anemones, the Town Council should come in state to see them. This was esteemed a great honour; and the Mayor managed in his progress to drag the skirt of his velvet robes over the bed of Anemones. The feathery seeds adhered as a matter of course; the blooms of these flowers coming in such long succession and the fruit ripening fast, so that flower and fruit were clustered together; and no sooner did he reach home than with most unlordly haste he picked off the seeds and sowed them. The next year his garden was adorned with the coveted flowers!

The Adónis (A. autumnális, or Pheasant's-eye, *Plate I.*, *fig.* 4), stands next in order to the Anemones; its little crimson flower gleaming from a miniature thicket of interlacing leaves. These leaves are frequently subdivided and very narrow, and the crimson cup is too small to be showy; it frequents corn fields, and is a constant ornament of old-fashioned gardens.

We now come to the true Ranunculus family, the head of the clan, called in English Crowfoots. Here the sepals and petals are both present, their number varying from three to six; the carpels are numerous, and end in a kind of horn. We have already noticed the poisonous nature of the whole order; this family of it possesses the quality in an especial degree. Old Gerarde thus writes of it: "There be divers sortes or kindes of these pernicious herbes comprehended under the name of Ranunculus, or Crowfoote, whereof most

---

* Varieties of this and other species are described by Tournefort as studding the fields in the islands of the Archipelago.

are very dangerous to be taken into the body. The chiefest vertue is in the roote, which, being stamped with salt, is good for those that have a plague sore if it be presently tied to the leg, by means whereof the poison and malignitie of the disease is drawn off from the inwarde partes, for it presently raiseth a blister to what part soever of the body it be applied. Apuleius saith further, that if it be hanged in a linen cloth about the neck of him that is lunatic, in the waine of the moon, when the signe shall be in the first degree of Taurus or Scorpio, that he shall forthwith be cured." Such was the strong admixture of ignorance and superstition which tainted the truth three centuries ago.

There are a great number of species of Crowfoot. The Creeping Crowfoot is the common Buttercup; and the bulbous and acrid species also receive the name of Buttercup (Ranúnculus bulbósus). I cannot tell you in how many rambles I have found these plants; from the days when, just able to run, we held the golden cups under each other's chins to see "if we loved butter," to the present time each season has witnessed

> "The Buttercups across the mead
> Make sunshine rift of splendour."

When Fanny was searching for wild flowers in the Lizard district last June, she found the Hairy Crowfoot (R. hirsútus), quite abundant in the pastures; it resembles the Creeping Crowfoot, but is distinguished by its hairs and its rather smaller paler flowers. A Shropshire ramble furnished Edward with the Biting Crowfoot (R. scelerátus), which he found growing on the margin of ponds and ditches about Kemberton. I found it in similar situations in Cheshire some years ago, but it is by no means a common plant. The lower leaves are palmate, or cut in fingers like a hand; the upper have only three divisions. Both leaves and stem are thick and watery; the plant grows from one to two feet high, and the flowers are small and uninteresting. It is the most poisonous member of the family.

There is a Corn Crowfoot (R. arvénsis), which somewhat resembles this, but it is less clumsy in form, and its large prickly carpels are both curious and pretty. I have frequently found it in corn fields in Kent, Yorkshire, and elsewhere. Two of this family have simple, narrow leaves; they are called the Greater and Lesser Spearworts. The larger species is rare. A specimen was sent to me from near Darlington; and I have seen it growing freely in a pond in the Botanic Gardens at Edinburgh. It is a stately plant with long, spear-shaped leaves, and large golden flowers nearly as big as those of the Marsh Marigold. It is quite erect, and grows to the height of three feet: this is R. lingua. The Lesser Spearwort (R. flámmula), is common everywhere on boggy ground and ditches. In the former situation it grows half prostrate; in the latter, it becomes erect and very succulent. In this state it may easily be confounded with the Greater Spearwort, only that it is never hairy, and its carpels terminate in a mere point, while those of the larger species have a sword-shaped beak.

The last of the yellow Crowfoots is the Lesser Celandine (Ranúnculus ficária, *Plate I., fig.* 6), so well known and loved by the poet Wordsworth.

> "Ill befall the yellow flowers,
> Children of the flaring hours:
> Buttercups that will be seen
> Whether we will see or no;
> Others, too, of lofty mien,
> They have done as wordlings do,
> Taking praise that should be thine,
> Little humble Celandine!
>
> "Prophet of delight and mirth,
> Scorned and slighted upon earth!
> Herald of a mighty band
> Of a joyous train ensuing,
> Singing at my heart's command,
> In the lanes my thoughts pursuing;
> I will sing as doth behove
> Hymns of praise of what I love."

No wonder that this cheerful flower should be a favourite. It is one of the first that appears as a herald of spring; and few will look upon its glossy star-like petals without a thrill of joy at the mute announcement that the hard, cold winter is past, and the time of the singing birds and budding blossoms is come. The Lesser Celandine has more petals than the other members of its family, and undivided spade-shaped leaves and tuberous root; so it is well distinguished from the rest.

We have three distinctly marked white species:—one with hairy leaf and short stem, which frequents the summit of our highest mountains (Ranúnculus alpína); one with an ivy-shaped leaf and tiny flower, growing in water (R. hederáceus); it is a frequent ornament of ponds in Yorkshire, Cheshire, and elsewhere. The third is the prettiest and most frequent of the white species, covering ponds and the margin of lakes with its fragile white blossoms, while its stems extend many feet, bearing round five-lobed fleshy leaves on the upper part, which float like the flowers on the surface of the water, and thread-shaped much-divided ones on the lower part, which are always immersed (R. aquátilis, *Plate I., fig.* 5).

The little Mouse-tail (Myosúrus mínimus), with its spiked cluster of tiny flowers, claims a near relationship with the Buttercups, because of its many carpels and stamens fastened to the receptacle. It is an inhabitant of gravelly fields, but has never rewarded my research.

A much more natural relation of the Buttercups is the Globe-Flower (Tróllius europǽus, *Plate I., fig.* 7). Its petals are larger and of a paler hue, and more numerous than those of any of the Buttercups; they are all cupped, and meet in the centre so as to enclose the stamens and carpels in a perfect globe. We used to be very proud of it, as it grew in great luxuriance in a corner of a shady field bordering the road between Richmond and Reeth. There were but a few plants, but we felt them most precious. But last June we were making an excursion across the Swaledale Moors to visit one of the numerous lead mines with which those hills abound, we came on a subalpine pasture covered with the Globe-Flower as thickly as we constantly see the Buttercup spreading its golden blooms. Knives and pocket-handkerchiefs were quickly produced, and we had soon procured a bundle of roots to plant in our shrubberies. How proud we have felt ever since in the knowledge that our neighbourhood boasted such a profusion of the treasured plant! In this manner it is frequently found in the hilly districts of the north of England.

The Marsh Marigold (Cáltha palústris, *Plate I., fig.* 8), is a still more showy plant, its large golden blossoms and bright green, glossy, heart-shaped leaves forming a familiar and much-valued ornament of marshy land. Like the Globe-Flower, the Marsh Marigold is without calyx. Last year Edward was much delighted with the many garlands carried about for show by the poor children in a village in Kent, and the one to which he gave the preference was formed entirely of Marsh Marigolds and purple Orchises.

The Hellebores also belong to this order; and although their green flowers suggest no relationship to their golden companions, yet the form of the blossom and its many carpels, and the dark glossy hand-shaped leaves carry out a fair amount of resemblance as soon as the attention is directed to them.

The Green Hellebore (Helléborus víridis), grows in woods about Bedale in Yorkshire, where I have found it frequently. Like its brethren, the Stinking Hellebore and Christmas Rose, it has no petals, and the petal-like calyx becomes leafy, and remains till the seed is ripe. The plant grows to a foot in height. Fanny has specimens gathered at Hill Deverill, in Wiltshire.

The Stinking Hellebore she has found near Shaftesbury, in Wiltshire. It has a disagreeable smell, and the pale green flowers are tinged with purple (H. fœtidus). Both of these plants flower early in the spring.

The pretty white Christmas Rose, so called because of its flowering at that season, is also a Hellebore, and its calyx is a brilliant white. Its beauty and purity, as well as the season when it blooms, insure it being a favourite in every garden.

The little Winter Aconite, which blooms so very early in the spring, its large green sepals giving it the appearance of a little boy of the olden time with an ample frill under his chin, and the bright blue and pink and white Hepaticas, so charming in our gardens and in the Swiss woods at the same season, also belong to this order.

The Poisonous Baneberry (Actǽa spicáta), one of the rare ornaments of the hilly districts of the north, with its spike of pink-tinted flowers, more nearly resembling a Spiræa than a Ranunculus, represents another family of this order. Herb Christopher is another name for it. Its leaves are frequently divided, and its fruit consists of a poisonous berry.

The Columbine forms an agreeable contrast to the golden Ranunculuses; it is generally purple, though sometimes pink or white. Growing to the height of two feet, its stem slightly branched, and bearing drooping blossoms at the end of the branches, its appearance is very elegant. The sepals are purple as well as the petals, and these last are in the form of a pendant cornucopia. A fancied resemblance in the cluster of spurred petals to a nest of young pigeons procured it the name of Columbine. The upper leaves are three-parted, the lower ones twice three-parted. Familiar as this plant is in the shrubbery and garden, it is rare in a wild state. We were much delighted two years ago when informed by an eminent botanist that the Columbine was to be found beyond the Wickliffe woods. We started on a hot day in June, leaving Richmond very early, and searching every field after the aforesaid woods were passed. My companion found it while I was examining another field, and we considered our fatigue well expended in procuring wild specimens (Aquilégia vulgáris, *Plate I., fig.* 9). Twamly calls this "Folly's Flower," likening its cluster of spurred petals to a fool's cap rather than to a nest of young doves. He says—

> "Then gather Roses for the bride,
> Twine them in her bright hair;
> But ere the wreath be done—oh let
> The Columbine be there.
> For rest ye sure that follies dwell
> In many a heart that loveth well.
>
> "Gather ye Laurels for the brow
> Of every prince of song!
> For all to whom philosophy
> And wisdom do belong;
> But ne'er forget to intertwine
> A flower or two of Columbine.
>
> "Weave ye an armful of the plant,
> Choosing the darkest flowers,
> With that a wreath blood-dipped bring
> The devastating powers,
> Of warrior, conqueror, or chief;
> Oh twine that full of Folly's leaf!
>
> "And do ye ask me why this flower
> So fit for every brow?
> Tell me but one where folly ne'er
> Hath dwelt, nor dwelleth now,
> And I will then the Laurel twine
> Unmingled with the Columbine."

The Monkshood and wild Larkspur are nearly allied to the Columbine. Like it they have purple flowers, though the blue predominates more in them and the violet in the Columbine. The Monkshood has no calyx; its upper petal is hooded, hence its English name: and its two side petals are hairy on the outside. My specimen was found in an old quarry near Gravesend, and I saw it frequently in groves in Switzerland; but it can hardly be accounted a wild plant. It is very poisonous, especially the roots, which sufficiently resemble Horseradish for some fatal mistakes between them to have occurred. A habit of one part of the root dying off and a new part forming and throwing up branches causes the plant to change its position perceptibly, so that in a few years it has moved several inches. The seeds are accounted very dangerous to the eyesight, and many parents will not suffer it in their gardens for fear of the small seeds being blown into the eyes of the children. A medicine is formed from the Aconite poison; but it is chiefly used externally. Its beneficent effects are best understood by the homœopathists: in their hands it allays fever, procures sleep, and thus alleviates much human suffering (Aconítum napéllus, *Plate I.*, *fig.* 10).

The Larkspur (Delphínium consólida), is occasionally found in corn fields. It is also without a calyx, and has five petals, and a nectary containing poisonous honey. Its form is light and graceful, and its leaves narrow and much-divided. There are many handsome species of Larkspur cultivated in gardens —one with a single crowded spike and double blossoms, the colour of which was purple; pink, grey, or white used to be much in vogue, and, planted in lines, always reminded me of Lombardy Poplars. The rich blue species now in fashion for ribbon-borders, &c., are perennial, and far exceed in beauty those prized in years past.

But the king of the order is the wild Pæony (Pæónia corállina), the stately ornament of the island called the Steep Holmes, in the mouth of the Severn. Fanny would have got a specimen if it had been possible, and to that intent she was eager to join an excursion from Clevedon to the island in question, the only known habitat of this noble plant; but great was her disappointment when she found that the owner strictly forbade the gathering of a single flower! So I fear the crimson beauty must continue absent from our collection. Our gorgeous garden Pæonies are most of them mere double varieties of this hero of the Steep Holmes, and the tree Pæony is another species of the same family. In the Crimea, Hooker states that the seed of the Pæony is surrounded by a red pulp, the juice of which affords a beautiful purple dye.

This family completes the Ranunculaceæ order.

CHAPTER II.

NYMPHÆÁCEÆ—PAPAVERÁCEÆ—FUMARIÁCEÆ—BERBERIDÁCEÆ.

> " We are the sweet flowers
> Born of sunny showers,
> Think whene'er you see us what our beauty saith;
> Utterance mute and bright
> Of some unknown delight,
> We fill the air with pleasure by our simple breath.
> All who see us love us;
> We befit all places;
> Unto sorrow we give smiles, and unto graces graces."
>
> LEIGH HUNT.

THE second order of Thalamiflorals has but few British representatives; but among these is the most attractive of all our wild flowers, the white Water Lily. Distinguished from the Ranunculaceæ by its long-enduring calyx and carpels with solid partition, the members of this order are like their predecessors in their numerous stamens and pistils. The first white Water Lilies I ever saw were in a pond near Copgrove, in Yorkshire. I was but a little child, but I stood rapt in amazement. Had I read Walter Scott's poems, I might have described the flower in his words:—

> "The Water Lily to the light
> Her chalice reared of silver bright."

As it was, I only conceived an absorbing desire to possess one. They were quite beyond reach, and I could not realise my desire; and I remember that for years the possession of Water Lilies mingled in every airy castle that I built. Five years ago I was with a pic-nic party at a pretty lake in Cheshire. We were all near relations—brothers and sisters, and cousins scarcely differing from brothers and sisters. Again I beheld the beautiful white Water Lily, and by its side the somewhat less attractive yellow one. I exclaimed at the sight of them, and said how long I had wished for them; then I began to aid in spreading the cloth for our entertainment. Turning to the lake again I was terrified to behold a young cousin standing up to his hips in water, and coaxing the Lilies towards him with a long stick. They often eluded him, for their stalks, six or eight feet long, acted as cables, and the flowers ducked under water as he tried to draw them, and reappeared at a greater distance from him, as if they were endowed both with locomotion and reason. But he succeeded at last, and I became the proud possessor of a handful of Lilies, with their suitable accompaniment of broad floating leaves (Nympháea álba, *Plate II.*, *fig.* 3; Núphar lútea, *fig.* 4). Edward has beautiful specimens from the Avon, near Warwick, and Fanny from the neighbourhood of Clevedon.

The superb Victoria Regia, which has justly attracted so much attention in the hothouse at Kew, is a member of this family. Its flower resembles that of our white Lily, except in being larger and tinged with pink, and its leaves are so large that a child may safely stand on one of them.

The third order is that of the POPPYWORTS, the distinguishing features of which are a calyx of two sepals, which fall off soon after the flower opens, numerous stamens, and milky juice: they have generally four petals. The familiar red Poppy, at once the ornament and bane of corn fields, is well known to us all. It is difficult to say whether its effect is most beautiful among the green or the ripe corn. The seed-vessel is round and black, and divided into chambers; the stamens are purplish-black, and the leaves are deeply cut, so as to be divided into sharp segments. It is the fashion in general to turn a moral lesson to the disadvantage of Poppies, regarding them as emblems of vanity and worldliness; but I would look on them in a more kindly light, as examples of the fact that beauty is present everywhere, and is nowhere without its charm (Papáver rhǽas, *Plate II., fig.* 6).

> "See the merry Poppies, all amid the waving corn!
> Peeping up with blushing face to greet the cheery morn
> Here and there, and everywhere, their scarlet hue is seen,
> Always dotting, spotting, blotting, o'er the surface green.
>
> "Poppies! pretty Poppies! to me you seem to sing;
> 'Tis not to my eye alone that pleasure true you bring;
> In my ear I seem to hear your gentle whisp'ring voice
> Softly crying, murm'ring, sighing, 'Sons of men, rejoice!'
>
> "Sing on, sing on, brave flow'rs! I lend a ready ear,
> Chant high your tuneful melody, I'm all attent to hear;
> And, list'ners kind, if you're inclin'd the Poppies tale to learn,
> I'll tell it you in language true, if not in 'words that burn.'
>
> "'Life is like a furrow'd field,' I hear them softly say,
> 'Broadcast sown with cares and griefs, which spring up day by day;
> 'But ever there, 'mid crops of care, some bright-hued joy appears
> 'To teach you men to hope again, for smiles amid your tears?'"

The Long-headed Poppy is smaller than the common one; its petals are of a lighter red, and its seed-vessels are elongated. I found it as a garden weed about Warminster, in Wiltshire. There are a rough-headed Poppy with a bristly seed-vessel, and a smooth-headed one with yellow anthers; but I have not found either of these. The White Poppy (P. somníferum, *Plate II., fig.* 5), is the most important member of the family, because it produces the opium of commerce. This is obtained by cutting the nearly-ripe seed-vessels; the juice which flows from the incision hardens and forms opium. Laudanum and morphia are preparations of this. Every one knows the kind of effect which these drugs produce. Sometimes they are to be regarded as a blessing, soothing the agonies of pain, and compelling sleep where the brain is overpressed; but the snare of using them too freely is a temptation ruinous to so many, that we must regard the lulling juice of the Poppyworts as more dangerous than the biting sap of the Ranunculaceæ. Leigh Hunt makes the white Poppies introduce themselves winsomely :—

> "We are slumbering Poppies,
> Born of Lethe hours,
> Some awake, and some asleep,
> Nodding in our bowers."

This sounds very innocent; and, as long as we can picture the sleep to be that of the exhausted sufferer, we give the Poppies all honour; but from the "Lethe hours" of the opium-eaters' Fool's Paradise we turn away in sorrow and disgust. This plant is frequently cultivated in England and its heads preserved for fomentations; in India, Egypt, and China it is very extensively grown as an article of commerce. The leaf is smooth and of a bluish-green or glaucous hue. These specimens were gathered wild near Clevedon.

The yellow or Welsh Poppy (Meconópsis cámbrica), is a native of Wales; but I have found it in the north of Yorkshire. It is yellow; the calyx drops as the blossom opens, and often falls into the cup formed by the corolla. The stigma and anthers are also yellow, and the leaf is much-divided and hairy.

The Horned Poppy (Gláucium lúteum, *Plate II.*, *fig.* 7), is another near ally. Its leaf is fleshy and of the same colour as that of the white Poppy, and, like it, it clasps the stem. The distinguishing feature of the plant, and that to which it owes its name, is the elongated horn-shaped carpel. Its petals are of a fine yellow. I have found it near Hastings, and Edward has specimens from the Lancashire coast. There are a purple Horned Poppy and a crimson one; but they are very rare, and we have none of us been able to procure specimens.

The greater Celandine (Chelidónium május, *Plate II.*, *fig.* 8), represents the last family of the Poppywort order. It has a pod-shaped seed-vessel like the Horned Poppy, two sepals, four petals, and a bright green leaf, deeply lobed and pinnate. The juice is orange instead of white, and it has a burning quality, so that it has been used to cure warts with. Edward found it growing under an old wall near a village in Kent, and I have seen it in great abundance about Dornington and Stoke Edith, in Herefordshire, and elsewhere in similar situations. The yellow flowers are not large enough for the cluster to be gay, so the very fresh green of its foliage is its greatest recommendation.

The small order of FUMEWORTS comes next to the Poppyworts. We made an excursion last summer to see the magnificent ruins of Fountains Abbey. There, in the garden of the old monks, as well as in wild spots among the ruins, was growing the yellow Fumitory (Fumária lútea), as I remembered to have seen it growing when I was a little child, and as it probably grew centuries before. These plants have two sepals, which fall off when the flower opens, and four petals, one or two of which have little bags at the base, which seem like a spur at the back of the flower. The flowers are arranged in an irregular inclined spike, and light much-divided leaves grow upon it in pairs. The Bulbous Fumitory is called a wild plant; but I have only seen it in old gardens. It has a bulbous root and purple flowers.

The common Fumitory (Fumária officinális, *Plate II.*, *fig.* 9), is a familiar garden weed. The glaucous, much-divided leaves resemble so closely those of the brilliant orange Eschscholtzia that the young plants are often spared in mistake for the welcome seedlings. Its flower is white, just tipped with pink and green. Altogether it is very inexpressive.

The Ramping Fumitory (F. claviculáta), is the prettiest member of the family. Never shall I forget its graceful charm as I first beheld it from the top of a coach passing through the Trossacks. It was September, and we had wandered in vain among the lovely scenery round Callander in search of alpine plants. I had sorrowfully resigned myself to returning without floral mementoes of that region of poetry and beauty, and in this spirit I had ascended the coach to pass forward to the Trossacks. A fellow traveller was reading aloud "The Lady of the Lake," the Brig o' Turk was passed, and my eyes had rested from the vain search for flowers, and were feasting on Loch Achray and the hills beyond, when I espied a dainty garland of tender green mixed with clusters of pale wax-like blossoms. A humane friend procured me a quantity of the plant, which proved to be the Ramping Fumitory. I know of no medicinal or useful properties in these plants.

The fifth order is that of the BERBERIDS (Bérberis vulgáris, *Plate II., fig.* 1). We have but two British families in it. Sepals, from three to six, falling off; and petals either the same or double the number; stamens equalling the petals in number, and anthers opening by valves, and so discharging the pollen. The Berberry shrub, so ornamental whether in flower or fruit, is the familiar representative of the order. It grows freely in some of our Yorkshire woods, and in many other counties. The pendant clusters of yellow flowers appear in May, each bunch guarded by a three-pronged thorn. The stamens are curiously sensitive; if you touch them lightly at the base with a straw or a pin the stamen contracts, and the anther bends forward and strikes against the stigma. The strange sensitiveness is probably a provision of nature, securing that if any insect walk round the flower each anther touches the stigma, the collision probably causing the valves to open, and thus discharging the fertilising pollen. This shrub is liable to a tiny fungus, which appears in clusters like an orange blot on the leaves. In former days it was thought that the Berberry blight gave rise to smut and bunt, the destructive diseases of corn; and farmers would not suffer the Berberry in their hedge-rows. The berries make a fine preserve, and, when merely kept in salt, they form a pretty garnish for savoury dishes.

The Barrenwort (Epimédium alpínum, *Plate II., fig.* 2), is the other British member of this order. My specimen was gathered near Frome, in Somersetshire; but I am told that it is no longer found there. It may still be occasionally met with among the Pentland Hills, near Edinburgh. It has four petals of a crimson colour, four hollow yellow nectaries lying upon the petals, which form the figure of a Maltese cross, and four stamens. The foliage is very handsome, the leaflets are in clusters of nine, they are slender, heart-shaped, bright green, and fringed. The kind, genial old botanist who first introduced me to this plant said he called it the "Happy Medium," in fanciful allusion to its botanical name.

CHAPTER III.

CRUCÍFERÆ.

" ' Not to myself alone,'
The little opening flower transported cries,
' Not to myself alone I bud and bloom ;
With fragrant breath the breezes I perfume,
And gladden all things with my rainbow dyes.
The bee comes sipping, every eventide
His dainty fill ;
The butterfly within my cup doth hide
From threatening ill.' "
SARGENT'S *Collection.*

THERE is no family of plants whose uses are wider spread and who live less "to themselves" than the CRUCÍFERÆ or CRESS-WORTS, the sixth natural order. Their simply-constructed flowers utter God's praises with their "silent lips," showing forth His wisdom in their easily-recognised construction, and His love in scattering so suitably and abundantly these most useful and healing herbs.

The cruciform flower, with its four petals, four sepals, four longer and two shorter stamens, and the entire absence of any floral leaf or *bract* at the juncture of the flower-stalk with the main stem, are distinguishing marks which never vary, and which are open even to a child's observation. The uniformity and simplicity of this rule becomes a matter of no small importance when we consider that *all* the plants of this order are wholesome, and many afford agreeable food, either cooked or eaten as salad. The commonest sea-side plants, extending into very high latitudes, are the Scurvy Grasses, the most valuable antidote to that frightful scourge of seamen. When in Anson's famous expeditions this disease was consuming the power of the crew, the doctor dare not allow them to eat any vegetable except grass when they landed on unknown shores.

How valuable would this simple rule of botany have been which makes the cross-shaped flower and bractless stalk an ensign of safety, an invitation to eat and live! Surely, without too fanciful a straining of metaphors, we may see here an indication of a more holy mystery, a greater benefaction to humanity, where the cross is raised over all created things, whether upon the earth or above the earth, and by the power of Him who hung upon it the curse of sickness and death is taken away, and health of the entire nature, both physical and spiritual, established for eternity.

The Cressworts have two different kinds of seed-vessels, a pouch and a pod. The tiny Rock Hutchinsia stands first in the pouch group; it is a fairy-like plant with pinnate leaves, and a spike of minute white flowers, its whole stature not exceeding from an inch to an inch and a half. I cannot say that I found it wild, for it was growing upon the wall of a botanic garden; but Mr. Ward, the proprietor, had taken the seed from wild specimens.

The same botanist furnished me with a specimen of the Penny Cress (Thláspi arvénse), which I afterwards found in abundance in corn fields near Ross, Herefordshire. Its flowers are also insignificant; but its large pouches are very striking. There are an alpine and a perfoliate Penny Cress, but I have not found either.

The Shepherd's Purse (Capsélla bursa-pastóris), is a well-known weed; its compact spike of heart-shaped pouches forms a neat object, and adorns the dusty road-side and neglected pavement, as well as the borders of fields and lanes.

The Teesdália is a plant as small as the Hutchínsia, with a leafless stalk and entire leaves. We have none of us been able to procure specimens.

The Pepperworts are a large family. The Broad-leaved Pepperwort (Lepídium latifólium), is a wand-like plant. It was sent to me from Scarborough, where it flourishes on swampy ground.

Fanny found the Hairy Pepperwort (L. hírtum), growing freely on the Chough Rock, one of the many beautiful cliffs adorning the coast near Looe, in Cornwall. The situation is very much exposed, commanding a wide prospect of sea and coast; and, if ever little plant needed its warm coat of hairs, this modest little Cornishman required his. It is a curious fact that the more cold and bleak the situation, the more hairy does the vegetation become; while, if the same plants are removed to a sheltered locality, they gradually become smoother —a wonderful example of the providential adaptation of every atom in the machinery of creation.

Near the Lizard Point Fanny found the English Scurvy Grass (Cochleária ánglica, *Plate III.*, *fig.* 1), growing on rocks from which issued tiny streams, supplying the plant with moisture enough to nourish its succulent leaves and stem. This plant grows on all our coasts and on the banks of our salt rivers, its purple-lined leaves and clusters of pure white flowers forming a very pretty object. In this family the petals are no longer insignificant as in the Pepperworts, but of some size and of brilliant whiteness. The leaf of the one in question is ovate, stalked when arising from the root, but seated on the stem.

The Danish and Greenland Scurvy Grasses are also found occasionally on our shores, but are more abundant in the lands from which they take their respective names.

The Common Scurvy Grass (C. officinális), is abundant on the banks of streams, inland as well as seaward. It grows in great luxuriance on the banks of the Swale, making patches among the sward of such brilliant whiteness as to rival a flower-bed in an Italian garden. Its leaves are roundish, and its pouch of the same form, not veined as in the English Scurvy Grass.

The Horseradish belongs to this family; but though called wild, I must confess that whenever I have found it there were strong grounds for believing it to have been thrown out from a garden. While the foliage is the part used in the other Scurvy Grasses, the root is the edible portion of this member of the family, and we are all familiar with it in its scraped form as an adjunct to roast beef.

Of the Water Awl-wort (Subulária aquática), the next family in the Cresswort order, which grows at the bottom of the Highland lakes, we have no specimen.

The Whitlow Grasses have lobed petals, but in other points resemble their predecessors. The little Vernal Whitlow Grass (Drába vérna, *Plate III.*, *fig.* 2), so common early in spring on old walls, is a familiar object. Its star-shaped cluster of simple leaves is hairy and its stem leafless, while the cluster of flowers is large in proportion to the minute bulk of the plant. About Edinburgh the old turf-topped walls are gay with myriads of this pretty little plant, and at midsummer I have found it decorating heaps of rubbish around the mouths of exhausted lead-mines on our Yorkshire moors. Its pouches differ from those of the Scurvy Grass in being quite flat and oval. The Yellow Whitlow Grass (D. aizoídes), is more shrubby in its growth, and its clustered spike is thicker and more crowded. It adorns the rocks in mountain glens, and was transplanted from them to the rockery from whence my specimen came. The Speedwell-leaved Whitlow Grass we have none of us found; but Fanny has brought the hoary species (Draba incana), as a memento of one of her rambles on the banks of the Looe river. The flowers are smaller and their footstalks shorter than in the other species; the main stem is tough and woody, it and the leaves are covered with a mealy bloom. It grows a foot and a half high, and flowers in June.

Here is the purple Sea Rocket (Cakíle marítima, *Plate III.*, *fig.* 3), with its bright green fleshy foliage and thick oval pouches. I gathered it on the coast of Durham, near the little village of Roker, one June day. Young shoots of this plant and of a Sisymbrium are eaten in the Crimea as salad herbs.

The Sea-kale (Crámbe marítima), so familiar in our gardens, grows wild abundantly on the cliffs on the south coast, making a great show with its large spikes of white flowers and glaucous foliage. The edible species of this Cresswort order are all rendered milder, sweeter, and larger by cultivation. Though wholesome in the wild state, the increase of the sugar-like quality makes them more nutritious. Thus by diligent use is the excellence of God's gifts enhanced.

The Greater Wart-cress (Senebiéra corónopus), I found one scorching July day growing luxuriantly in some cracked and parched waste clay land near Norwood, in Surrey. The stems are tough and trailing, the leaves much-divided, and the flowers very insignificant. When Fanny arrived with her mother at Plymouth last June, and they were disappointed at finding no carriage to carry them into the country, they went to walk for half an hour on the East Hoe, from whence they got a beautiful view of the famous breakwater, and the richly-wooded rising ground to the right. There Fanny had the pleasure of finding this specimen of the Lesser Wart-cress, which was growing abundantly among the rocks. The heart-shaped seed-vessels and smaller leaves distinguish it from the larger species (S. dídyma).

She found the Sweet Alýssum (A. marítimum), at Clevedon. Its close short spike, hoary foliage, and sweet scent distinguish it from its allies. It is a common garden flower; and a variegated species is much in vogue at present for ribbon-borders, enclosing masses of scarlet Geranium, blue Larkspur, &c.

The Bitter Candytuft (Ibéris amára), is also found near Clevedon. Its flowers grow in a cluster rather than a spike, those towards the edge larger than those within. The brilliant white of the crowded cluster makes it a gay plant. It closely resembles the purple and white Candytuft so often seen in our gardens as an annual. The perennial Candytuft is a still more desirable border flower. Another familiar garden plant nearly allied to these, and belonging to the same order, is the purple Honesty (Lunária annua). There is a popular superstition that wherever this plant flourishes the cultivators of the garden are exceedingly honest. It would indeed be well if this analogy between the external and internal honesty held good always, for the former is a hardy plant, and there are few circumstances under which it will not flourish! The Woad (Isátis tinctória), with the juice of which the ancient Britons used to paint their bodies, is a member of this tribe; but I have not been able to procure a specimen of it.

We will open the Pod group with a Kentish treasure of Edward's. The Coralroot (Dentária bulbífera, *Plate III.*, *fig.* 4), is a very remarkable plant, the only one indigenous to Britain which bears germinating buds upon the stem. This habit is familiar to us as displayed in the Tiger Lily, whose stem-buds are produced in abundance, and by them the plant may be increased with enormous rapidity. Another beautiful peculiarity of the Coralwort is the underground stem, by some persons considered a part of the root. This is branched, with thick scales at the joints, pure white, and in general effect resembling coral—hence the English name. The fibres constituting the true root are attached to this coral-like stem. The plant is rare, it grows from one to three feet high, and its lilac blossoms resemble those of the Cuckoo-Flower. The leaves are of three leaflets in the middle of the stem, of one towards the top, and of five or seven where they spring near the base. Although I have called it a Kentish specimen, I see it is in fact a Sussex one, for Edward found it on moist ground in some plantations on the estate of Lilles Den. The stream which divides Kent from Sussex traverses this property, and the Coralwort favours the Sussex side of the stream. The Cuckoo-Flower (Cardamíne praténsis, *Plate III.*, *fig.* 5), so familiar an ornament of moist woods and fields, is abundant everywhere. The colour of the flowers varies to every shade between white and full lilac, and the leaves are *pinnate*—that is, composed of two rows of leaflets.

The Hairy Bitter Cress (C. hirsútum), is welcome in early spring for the fresh green of its foliage, although its little white blossoms are very insignificant. The Narrow-leaved Bitter Cress (C. impatiens), I found in the Chase Wood, near Ross, Herefordshire, growing by the side of the path in damp places, among a forest of the Wood Spurge. I greeted it joyfully, having never before seen any but dried specimens which had

D

been given me by Mr. Parry, of Warwick. It grows there in great abundance, the stems rising to the height of a foot and a half. The flowers are insignificant, often wanting the petals altogether, but the abundance of finely-cut bright-tinted foliage makes the plant attractive. The leaflets are narrower in this species than in the others.

The Marsh Bitter Cress (C. amára), is a very handsome plant. I have gathered it in the Ironbank woods, near Richmond, and also in Swaledale. The foliage is more abundant and of a lighter green than that of its sister, the Cuckoo-Flower. The spike is very compact, and the large brilliant white of the petals is relieved by violet anthers. The whole family deserve their name of Bitter Cress; but this species has the biting principle in a greater degree than the others, so as to be quite unpleasant to the palate. They all flower in April and May.

Swaledale has furnished me with two species of the Rock Cress, the characteristics of which family are oblong petals, and narrow pod with flat valves. The common Rock Cress (Árabis thaliána), grows on crumbling rocks in woods about Reeth. The foliage is hairy and slightly toothed, the height varies from one inch to half a foot.

The Hairy Rock Cress (A. hirsúta), is woody and very erect, often above a foot high. The root-leaves grow in the form of a star, and then the stiff stem shoots up from their centre, and the toothed leaves are seated on the stem at short intervals. It grows abundantly on old walls in Swaledale, flowering in May a little later than the last species.

Fanny found the Bristol Rock Cress (A. strícta), on Cadbury Hill, near Congresbury, in Somerset. Its larger flowers, with a cream-coloured erect petal, distinguish it from the other species. It blooms in the same month.

The Alpine Rock Cress (A. alpína), is not found wild, but it is a common garden flower. Its heart-shaped toothed leaves of white powdery texture and brilliant clusters of white scented flowers make it deservedly a favourite.

The Tower Rock Cress (A. turríta), is found on old walls about Oxford and Cambridge. Edward will, doubtless, procure it for us in a year or two. Its flowers are sulphur-coloured.

The Tower Mustard (Turrítis glábra), is a wand-like plant with numerous slender pods pressed close to the stem, and yellow flowers. We have none of us found it.

The Winter Cresses, or Wild Rockets, are a gay family of this order, adorning many a sandy bank and rubbish-ground with their spikes of golden flowers.

The common Yellow Rocket (Barbaréa vulgáris), is frequently found in such situations. It has a tough smooth stem, leaves of a full bright green, and of the form called *lyrate*—that is, not divided to the stem so as to deserve the term pinnate, but too much divided to be merely notched.

The Early Winter Cress (B. præcox), has smaller flowers and the stem-leaves are pinnate. It is not a common plant.

We now come to the true Cresses, or Nastúrtiums, in no way allied to the flower called by that name in our gardens, which is merely the namesake of these wild plants because of a similarity in flavour and in adaptation for salad. The Water Cress (N. officinále), is more generally known by its foliage than its flower. Few are unfamiliar with the fresh branches served up for breakfast salad, alike in town and country; but many would not recognise the spike of white cruciform flowers as belonging to the same plant. We associate much of pleasing romance with the idea of "Water Cress girls" going out from the smoky town to cull the fresh herbs at sunrise; but I saw the unromantic system of the trade in full last year, and my ideas are considerably tamed down in consequence. Travelling from London into Hertfordshire by an early train, I perceived acres of ground here and there, for miles along the banks of the New River, interspersed with broad shallow ditches. These had been cut on purpose to cultivate Water Cresses for the London market. Across these Water Cress gardens planks were thrown; and, in lieu of pretty maidens in blue petticoats and broad hats, with baskets, and garlands, and air-flushed cheeks, stooping in graceful attitudes to gather the aromatic herb, I saw men and boys lying on the said planks and picking hard at the Cresses as if they were paid at so much a bushel. When they arose their faces looked puffed and apoplectic, and they moved their planks and lay down again to work in a most matter-of-fact fashion. So much for the prestige!

The Creeping Yellow Cress (Nastúrtium sylvéstre), is a much smaller plant, with dark green pinnate leaves. It flowers from July to October. I gathered it some years ago near Little Ouseburn, in Yorkshire, in a clayey pasture.

The Amphibious Yellow Cress (N. amphíbium), is one of Edward's beloved Avon plants. It grows to the height of several feet, and its yellow flowers contrast well with the spikes of the purple Loosestrife, and the crimson umbels of the Flowering Reed. The leaves vary in form from simple oblong to pinnate; it flowers in autumn.

The annual Yellow Cress (N. terréstre), is less handsome, its petals being shorter than the calyx; it is found in Kinghorn Loch, in Fifeshire.

Perhaps the least interesting family, at any rate in appearance, of this order is that of the Hedge Mustard. The common Hedge Mustard (Sisýmbrium officinále), with its very insignificant sulphur-flowers, pods pressed close to the tough stem, and general dusty appearance, seldom tempt any but a botanist to gather its spikes.

Its sister, the Flixweed (S. Sophía), is a much more agreeable object. Its narrow doubly-pinnate dark green leaves and crowded spike of small yellow flowers are almost pretty.

The Broad Hedge Mustard or London Rocket (S. írio), is interesting as having appeared in great abundance after the fire of London. Some botanical books still refer you to "waste ground about London" for this plant, and I once vainly hoped to get it from a London correspondent. He repaired to a nurseryman for the desired plant, and received for answer that

"Yellow Rockets were quite out of fashion, and only double purple and white in vogue."

The Treacle Mustards are handsome plants. The Garlic Treacle Mustard, or Sauce-alone (Erýsimum alliária), would be valued if its strong scent of garlic did not warn off all contact. Its flowers are large and white, and its heart-shaped toothed leaves a bright glossy green.

The Worm-seed Treacle Mustard (E. cheiranthoídes), grows higher than the last species, attaining a stature of three feet. Its flowers are yellow, and its leaves lance-shaped. It is found in osier holts. My specimen was sent to me from the neighbourhood of Ely.

The Hare's-ear Treacle Mustard (E. orientále), has whitish flowers and oval leaves, and frequents sea-side places; but I have never seen a specimen.

Certainly the queen of the Cresswort order, as regards beauty, is the Wallflower (Cheiránthus Chéiri, *Plate III.*, *fig.* 6). They grow in profusion upon the magnificent ruins of Richmond Castle. The building stands upon a steep rocky hill, with the turbulent Swale flowing at its foot. The old walls are difficult to scale, and the gorgeous orange and umber Wallflowers flourish undisturbed in the crevices.

> " Flower of the solitary place!
> Grey ruins' golden crown,
> That lendest melancholy grace
> To haunts of old renown.
> Thou mantlest o'er the battlement
> By strife or storm decayed,
> And fillest up each envious rent
> Time's canker tooth hath made."
>   MOIR.

I once rashly mentioned my desire for some of these plants to a generous adventurous young midshipman. He said nothing at the moment, but the next morning he brought me both plants and flowers. Often as I have looked at those dangerous old walls I have shuddered to remember the risk he ran. One slip of the foot, and the daring which has since been a safeguard to his country and to humanity might have extinguished the abounding young life or crippled its powers for years!

If the Wallflower be the Queen of Beauty, surely the Sea Stock (Mathíola incána, *Plate III., fig.* 7), should be a leading star of her court. Its crimson petals and powdery foliage present a very pleasing contrast, and we owe to it the various beautiful Stocks which adorn our gardens. It is fragrant, though not so much as the Wallflower. My specimen was sent me from the Isle of Wight last May.

Prominent for utility, as the Wallflower and Stock for beauty, stands the Cabbage family (Brássica olerácea, *Plate III., fig.* 8). Yet, though no poet has ever thought of a rhyme in praise of either the plant or flower of the honest Cabbage, I have seen both arrayed in lovely attire. On a bright October morning I set off on an excursion to the Pentland Hills. There had been a white frost; and, as we approached the hills, we saw the rime still lingering on the shady side of the trees and buildings, while the country beyond was bathed in sunshine, and gorgeous with autumnal tints. In the "kail-yard" of a small farm-house rows of Cabbages, green and red, stood in prim order, each curled leaf fringed along veins and margin with Nature's diamonds. I stopped to admire, and thought how few hooped ladies in a ball-room could be so magnificently bejewelled, while their form resembled that of the plants before me so closely as to suggest the analogy. Fanny describes the maritime cliffs at Fowey, on the coast of Cornwall, as rendered gay with the yellow flowers and glaucous leaves of the Sea Cabbage, and the clusters of the red Valerian. If we imagine these plants relieved by dark rocks behind, and suspended over a crowded harbour and foaming sea, we can believe the beauty which they helped to form. By cultivation this Cabbage becomes finer, larger, and more nutritrious. It is the original of our divers cabbages.

Edward has nice specimens of the Wild Rape (B. nápus), with its glaucous leaves, heart-shaped at the base, and clasping the stem. Its flowers are paler than those of the Cabbage. The Isle of Man Cabbage is a pretty species, the petals veined with lilac. The Turnip and all kinds of Broccoli belong to this family. Rape seed is good food for birds, a useful oil is compressed from it, and the residue forms "oil-cake," a nourishing food for cattle, as effective as the "cod-oil" for the human species.

The Mustard family succeeds that of the Cabbage. The Charlock, or Wild Mustard (Sinápis arvénsis), is a great plague in corn fields. Its flowers are large and of a brilliant yellow, and contrast well with the rough lyrate leaves. It is as troublesome a weed as the Red Poppy, and scarcely less brilliant in appearance.

The common Mustard has four-sided pods, and the leaves are smooth. Edward has a specimen gathered in waste ground near Leamington.

I had the pleasure of finding the Narrow-leaved Wall Mustard (S. tenuifólia, *Plate III.*, *fig.* 9), near Norwood, one dreary November day. It is a slender, elegant plant, with a small spike of light yellow flowers, narrow stem-leaves, and lyrate ones from the root. The whole plant has a disagreeable smell. The seeds of the white Mustard are the most pungent, and consequently the most valuable.

The Radish family is the last in the Cruciferous order. Like its allies, the leaves are lyrate; but the jointed pods form a good distinguishing feature. The Wild Radish (Ráphanus raphanístrum), is Edward's trophy, brought from a piece of waste ground near Hawkhurst, in Kent. The petals are white or pale lilac, veined distinctly with a deeper shade.

The Sea Radish (R. marítimus, *Plate III.*, *fig.* 10), is Primrose-coloured, also veined. Fanny brought it from the beautiful cliffs near the Lizard Point. She has also specimens of the Dame's Violet (Hésperis matronális), the near ally of the Wallflower and Stock. It came from the home of the Bristol Rock Cress, Cadbury Hill, in Somerset. It has simple, lance-shaped, toothed leaves and pretty blossoms of a pinkish-lilac hue.

FORMS OF LEAVES.

We have here a nice collection of various shaped leaves. It will save much description if I show them to you now, and explain the terms used to express their various forms.

The Violet, the Twayblade, and the Lime (*fig.* A), are examples of *heart*-shaped leaves. That of the Marsh Pennywort (*fig.* B), Wall Pennywort, and garden Nasturtium serve

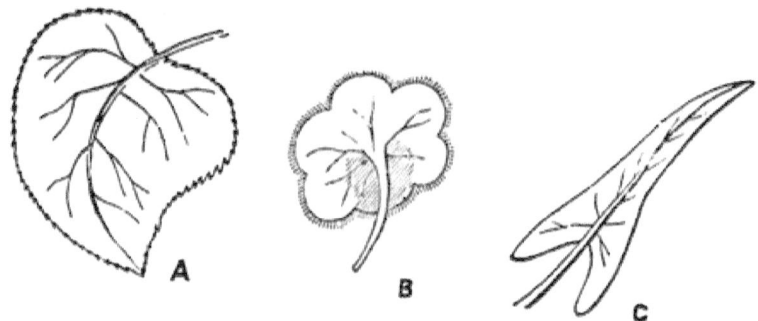

to illustrate the *buckler* shape. The common Sorrel (*fig.* C), and the Arrowhead have *arrow*-shaped leaves. The Bramble, Trefoil, Laburnum, and Wood Sorrel have *ternate* leaves (*fig.* D), or compound leaves composed of three leaflets. The

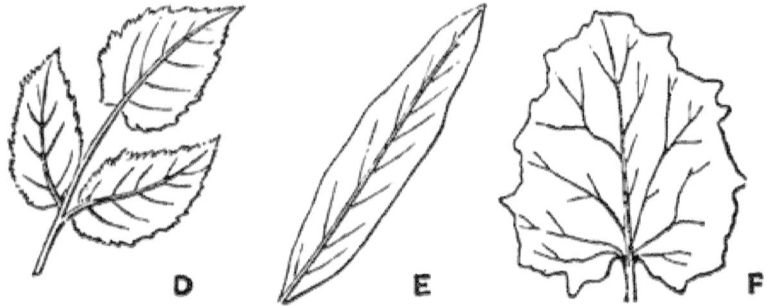

Cinquefoil has five leaflets and the Rose seven. A shape broad in the middle and tapering to the ends is called *lance*-shaped. The Wood Hawkweed, Deadly Nightshade, and Willow (*fig.* E), are our examples of this form. The Danish

Scurvy Grass affords an example of an *angular* leaf (*fig.* F); and the Ground Ivy (*fig.* G), Marsh Violet, and Lesser Celandine represent the *kidney*-shaped leaves. Of round leaves we have the Mint (*fig.* H B) and Wintergreen. In compound leaves

there is the simple *pinnate* form, which we have already noticed in the Cuckoo-Flower, &c. (*Plate III., fig.* 5); the *bi-pinnate*, where each leaflet is again divided; the *hair*-shaped pinnate and bi-pinnate, as the Fennel (*fig.* I); and the *tri-pinnate* still more divided. The grass-shaped leaf is rare, except in the second great class; but the Stitchwort affords an example among Thalamiflorals. The leaves of the Snowdrop, Crocus, Daffodil, &c., are all grass-shaped, modified in the form of their point to *acute, sword-shaped*, &c.; but these leaves have a different structure. We have already spoken of one broad mark of distinction between the first and second classes in the One or Two-lobed form of the seed. Another distinction equally clear and permanent exists in the venation of the leaf. In the Twolobed class the veins of the leaf form a network, the side veins running at almost a right angle with the midrib, and the lesser veins branching off in a similar manner from them. This *network* venation belongs to the Two-lobed class. In the Onelobed class the veins run parallel, starting from the footstalk and continuing in a line, more or less straight, to the point. This is plainly the case in all true Grasses and in the long narrow leaves of the bulbous plants. It is no less really so in other forms of the One-lobed foliage, as, for instance, the

Arrowhead, with its *arrow-shaped* leaves (*Plate XVII., fig.* 3), and the Lily of the Valley (*Plate XVI., fig.* 8), with its pointed *ovate* leaves. This distinctive character of the leaves is still more useful in deciding the class of a plant than the form of the seed, because the leaves are more generally present. A third point of difference exists, which can show the class of a plant even when the leaves are absent, and that characteristic is in the stem; but we will speak of that when stems are the subject of discussion. One remark more about leaves: In distinguishing species it is useful to observe the margin of the leaves. The simple pointed cutting on the edge of the Lime leaf (*fig.* A), is called *serrate*. Where the points are deeper and each point cut again, as in the Bramble (*fig.* D.), it is called *doubly serrate* (or sawed, like the teeth of a saw). Where little bits seem bitten out of the margin of a leaf, as in the Hawkweed (*fig.* K), it is called *toothed;* and where the cutting is simple but the edges blunt, as in the Mint (*fig.* H B), it is called *notched*.

Some leaves are haired. Hairs serve as a warm coat for the foliage, and they fulfil the further purpose of collecting the moisture from the air, and thus become doubly necessary for rock plants.

The leaves are the lungs of the plant, and when they are stripped off prematurely the plant sickens and often dies. To the human race leaves are very important—the shade they afford in summer, the salutary effect of them cooked as an article of diet, the refreshing tea, all these are claims upon our gratitude at the present day. In the old times, of which Varro speaks, when the leaves of Palms and Mallows had to serve the purpose of the convenient letter-sheet of to-day, and when the Sibyl wrote her oracles on dry leaves and then scattered them to the winds, they had a value of a different nature. In India a leaf still serves as a tablecloth, and the Jesuit missionary of Manilla dwelt in a hut formed of two Palm leaves.

CHAPTER IV.

RESEDÁCEÆ—CISTÁCEÆ—VIOLÁCEÆ—DROSERÁCEÆ—
POLYGALÁCEÆ—TAMARICÁCEÆ—FRANKENIÁCEÆ—
ELATINÁCEÆ—CARYOPHYLLÁCEÆ.

> "Now the heart is so full that a drop o'erfills it,
> Now we are happy because God wills it.
> We sit in the warm shade and see right well
> How the sap creeps up and the blossoms swell;
> We may shut our eyes, but we cannot help knowing
> That skies are clear and grass is growing.
> Everything is happy now,
> Everything is upward striving;
> 'Tis as easy now for the heart to be true
> As for grass to be green and skies to be blue,
> 'Tis the natural way of living."
>
> JAMES RUSSELL LOWELL.

CERTAINLY the influence of fine spring weather and of the beauty of spring flowers and foliage tends to produce cheerfulness and good humour; and when to these genial influences is added that which forms the mainspring of all—the recognition of the Creator in creation, and of His constant love and care, then the heart overflows with gladness and the "trees of the field" seem to "clap their hands."

The DYER'S WEEDS come next in our arrangement of Thalamiflorals. They are shrubby plants, with tough stems and long spikes of small flowers. A many-parted calyx, torn petals, and sessile stigmas are the distinctive features of the Rocket, or Dyer's Weed family.

The Common Dyer's Weed, Rocket, or Weld (Reséda lutéola, *Plate IV.*, *fig.* 1), is found commonly on waste chalky or limestone ground. Its leaves are bright glossy green and lance-shaped, its flowers greenish, and its tall spikes very narrow. This plant is much used for dyeing in France. The coloured paint called Dutch pink is made from it. Linnæus observed that its spikes always followed the course of the sun. Although it produces Dutch pink, yet the common colour of the dye formed from it is yellow. This specimen grew in a limestone quarry near Richmond, and I have often seen it in similar situations. It blooms in August.

The second member of the family, the Wild Mignonette (R. lútea), I brought from sandbanks in the bed of the Skell, near Ripon, in Yorkshire. It is more spreading in its growth, and its flowers are buff-coloured.

The third species is very rare (R. fructiculósa), the Shrubby Dyer's Rocket. Its broader spikes and straw-coloured flowers give it a close resemblance to the garden favourite, the fragrant Egyptian Mignonette (R. odoráta); but it lacks its scent entirely. My specimen came from Mr. Ward's botanic garden.

The family of the Cístus, or Rock Rose, succeeds that of the Rockets. These plants have five sepals, two external and prominent and three more concealed; five petals, very tender and easily falling off; and many stamens. They are generally shrubby plants with woody stems.

The common Cístus grows abundantly on rocky places throughout the kingdom; and its clear yellow wide-spread blooms are agreeably familiar to us all. (Cístus heliánthemum, *Plate IV., fig.* 2). The white Cistus I have never seen but in the neighbourhood of Clevedon, where it grows upon the grassy cliffs (C. polifólium). The Dwarf Cistus has much smaller flowers, of a pale yellow colour. It grows on the hills between Yorkshire and Westmoreland. Of the Spotted Cistus and Ledum-leaved species I have got no specimens. Gum ladanum is the resin of the Gum Cistus: it used to be collected in a very curious way. Goats were turned among the Cistus bushes to browse; they nibbled the branches, and the resin adhered to their beards and hardened there. It was then picked off for purposes of commerce. These bushes are still cultivated in Cyprus, but the resin is gathered with rakes.

Hooker considers this gum ladanum to be the myrrh ("Loth"), mentioned in Genesis xxxvii. 25, and to have been among the spices brought from Gilead to Egypt by the Ishmaelites. A crimson species which grows wild in Palestine is supposed to be the Rose of Sharon.

The VIOLET FAMILY has five sepals, five petals, five stamens, and one prominent hooded stigma. The ovary has three valves, and when it bursts it throws the seed with violence, so that it is scattered to a considerable distance. The Sweet Violet (Víola odoráta, *Plate IV.*, *fig.* 3), is a general favourite. My earliest remembrances are of "violetting" in fields about Wickliffe Lane and Littlethorpe, near Ripon, and of the joy we experienced as we gathered little baskets full of the purple and white varieties. They are much scarcer about Richmond, and are so seldom found in Scotland as to be termed there "English Violets." Fanny has a pink variety from Maiden Bradley, in Somerset, a county very prolific in this favourite flower. The tint is the same as that of the double Russian Violet which is cultivated in our gardens. The double purple one is a variety of the common Sweet Violet. These humble flowers have a wide range of habitation. They flourish at the foot of the Alps and in Arabia Felix; and Humboldt gathered them on the sides of the Andes and in the vallies of the Amazon. Napoleon Buonaparte adopted this flower as his emblem, and he was called "Père la Violette." Nor is the value of the Violet a fancy of recent origin; in the middle ages the prize awarded to the best poet was a golden Violet,—

"And in the golden vase was set
The prize—a golden Violet."

Milton depicts them as adorning the sinless earth, and Shakespeare honours them with his notice.

The juice of the Violet is used in medicine, and the plant is cultivated for that purpose. There are extensive Violet grounds near Stratford-on-Avon. This Violet juice forms the most delicate test for detecting acids and alkalies. The Romans used to make wine of Violets, and the Turks still infuse it in their sherbet. According to Lightfoot, the Highland ladies of former times used the Violet as a cosmetic, the old Gaelic receipt being, "Anoint thy face with goat's milk in which Violets have been infused, and there is not a young prince upon earth who will not be charmed with thy beauty."

The Sweet Violet has no main stem; its leaves and flowers spring from the roots. It increases rapidly by runners as well as by throwing out its seed.

The Hairy Violet (V. hírta), has a considerable resemblance to the Sweet one; but it lacks its scent, does not throw out runners, and its narrower leaves are covered with hairs, as well as the flower-stalks. I have found it near Richmond and in Durham; but it is not very common.

The Marsh Violet (V. pulústris), the third stemless species, flowers earlier than these, producing its paler blooms in March. I found it near Richmond in Yorks, and near Callander, and Fanny has leaves of it from near Harrowgate. Its leaf is kidney instead of heart-shaped, as in the species before mentioned.

The Dog Violet (V. canína), is a gay and frequent ornament of our woods and moors. It has a stem, and leaves and flowers grow from it, as well as from the root. It blooms later than the other Violets, is of a paler colour, with pencilled stripes across its white centre, and purple-tinged leaves.

The yellow Mountain Violet (V. lútea), grows freely on our Yorkshire moors, especially where water has stood. It is often seen growing side by side with the purple Butterwort, and the effect of the contrast is charming. In form it rather resembles a Pansy than a Violet; but both belong to the same family. The stem is very decided in this species.

The Corn Pansy (V. trícolor), is familiar as a corn weed, as well as a run-off plant in neglected gardens. Our handsome garden Pansies are, many of them, cultivated varieties of this; and, when left uncared for in a poor soil, they return to their natural state. Although this plant is called Tricolor (three-coloured), it is often entirely purple, and there is also a white variety. Here the stem is more decided than even in the Mountain Violet, and attains a length of several inches. The name Pansy is derived from the French motto, "Penzes à moi," and Shakespeare evidently has this in his mind when he says, "There's Pansies, that's for thoughts." Louis XV., when wishing to show favour to his physician Quesnay, devised for him a group of Pansies as his armorial bearings. Our English name, Heart's-ease, is a still prettier one.

We have a specimen of one Violet more, the cream-coloured Violet (V. láctea). It varies from the others in its pale colour, and leaves more lance-shaped than heart-shaped. I found this rare plant in a shaking bog near Tiverton, in Devon, some years ago.

The SUNDEW family has a three-valved seed-vessel and prominent stigma closely resembling that of the Violet, but the small regular petals and leafless stem, curved when young like fronds of Ferns, clearly distinguish the two. The leaves of the Sundews are fringed with long crimson hairs, bearing glands which collect the dew. They grow on footstalks, but lie in a starry form close to the ground. The tiny flowers are white and grow in spikes. Our English Sundews have five stamens; but in foreign members of the family there are often twice, thrice, or four times the number. When small insects touch the hairs, the leaf closes upon them and keeps them imprisoned. When pressed the leaves make a red mark on the paper.

The Round-leaved Sundew (Drósera rotundifólia), is a frequent denizen of our moors, luxuriating on the margins of peat bogs. I have found its crimson-tinted stars and slender spikes on the Yorkshire moors in very many places. I have gathered it on Rudd Heath in Cheshire; in Cornwall; and last autumn I found it in abundance on the swampy coast of Arran, and on the hills about Oban. Specimens of the

Greater and Long-leaved Sundews were sent to me from Leckby Carr, in Yorkshire. Their spikes of flowers differ scarcely at all from the familiar Round-leaved species, but the leaves are a long oval shape; and the Greater Sundew attains a larger size.

Another plant of this order is represented by our favourite, the Grass of Parnassus (Parnássia palústris, *Plate V.*, *fig.* 10); it is a widely-diffused plant, owing its name to its prevalence on Mount Parnassus. Swiss, French, and English hills alike offer it suitable homes among their swamps in the present day. It has five sepals, five snow white petals, five nectaries fringed with threads bearing globes, and tinged with cream colour. The leaves are heart-shaped, one only growing on the stem, the others springing from the root; it grows about a foot high. We frequently find it in damp places among our Yorkshire hills; and when on a tour in the Highlands I saw whole fields white over with its blossoms. It flowers from July to October.

Of the next family, that of the MILKWORTS, we have only one British representative; but that one is so frequent as to be familiar to us all, whether known by name or not. The great peculiarity in this plant (Polýgala vulgáris, *Plate IV.*, *fig.* 4), is that two of the sepals are wing-shaped and coloured, while the small petals adhere to the tube formed by the stamens, thus giving the flower a butterfly form, and inclining a young botanist to place it among the Papilionaceous order. The seed-vessel is pouch-like, heart-shaped, and pendulous. Near Flimwell, in Kent, I gathered three varieties of this plant upon one bank—blue, pink, and white. On our own moors I have found the same varieties, but not growing near together. The plant is abundant upon limestone rocks and I know of a nurseryman who collected a quantity of it last year for sale as a rockery plant.

We will introduce here the TAMARISK order, represented by Fanny's beautiful Tamarisk branches. In the bleak country of the Lizard district surrounding the village of Landewednack, she found a row of Tamarisk shrubs (Támarix gállica). The branches are beset with tiny glaucous leaves, smaller than those of the Larch, indeed seeming rather like scales than leaves; and they were relieved by spikes of small pink flowers, very pale and delicate, and boasting five petals and five stamens. This is the largest of our sea-side shrubs. It prevails about Sandgate and elsewhere on the south coast, as well as in the Lizard district. Its name is taken from the river Tambra, in Spain, formerly called Tamariscus, on the banks of which the shrub abounds. Forbes states that in Britain are found the remains of four distinct races of plants. Twelve species are found in the south of Ireland which belong to the Flora of Spain, and I fancy the Tamarisk must pertain to this group. The south-east of Ireland, and south-west of England, bear plants belonging to Brittany and Normandy; while the mountains of Scotland, Cumberland, and Wales exhibit a Flora resembling that of Switzerland, and more closely, of Lapland. The general type of the British Flora is chiefly analogous to the German. Many scientific men have supposed that the seemingly foreign plants have been introduced in ballast, but Professor Forbes was of opinion that each distinct group belonged to a different geological period, when a different climate prevailed. He considers the Spanish type the most ancient. The Arctic plants he ascribes to the "drift period."

The Sea Heath order (FRANKENIÁCEÆ), is distinguished by four or five sepals, five clawed petals and round anthers. The flowers are pale pink and very small. Lindley thus describes them:—"Little obscure plants usually inhabiting the neighbourhood of the sea, and of no importance to man."

The WATERWORTS (Elatináceæ), are still more insignificant and unimportant than the Sea Heaths. They have from three to five petals and sepals, and double the number of stamens. Our British species inhabits the margin of ponds and lakes (Elátine hydropíper).

E

We now come to a very extensive order of plants, the fourteenth in the British list—that of the Pink. It includes two tribes, the Pinks proper and the Chickweeds. The plants of this order have five sepals, united in some families, disunited in others, ten stamens, and a seed-vessel of one cell. In the first tribe the stamens join the stalk below the ovary, or seed-vessel; and the sepals are united. In the second tribe the stamens are joined in a ring, also below the ovary, and the sepals are free. Both tribes have simple leaves placed opposite each other on the stalks.

Mr. Ward gave me a specimen of the Deptford Pink (Diánthus arméria), the flowers of which grow in clusters and are spotted; they are quite scentless. The plant is occasionally found in waste places.

I have no specimen of the beautiful rose-coloured Proliferous Pink (D. prólifer).

I had much difficulty in getting this piece of the Clove Pink, or Gillyflower (D. caryophýllus). It grows upon the ruined walls of Fountains Abbey; but visitors are not allowed to touch the flowers. I imparted my great wish for a blossom of the Pink to a very spirited friend who was of our party. She was most anxious to aid me, and began a systematic attack upon the good nature of the guide. Both she and I had visited the place often enough to know all the descriptions by rote, but she suddenly became exceedingly interested in the old story. The guide was charmed to gain so attentive and intelligent a listener, and he presently began to regard her with great favour. Then she made her request, and at once received a number of the desired blossoms as a reward for her diplomacy. This species is the origin of all our garden Pinks from the plain rose and white ones which accompany the Mint and Southernwood in the cottage garden, to the delicately-bordered and variegated blooms which carry off the prizes at flower shows. The Latin name of this family, Dianthus, means "Gift of God," because of the estimation in which these fragrant flowers were held. In olden times the petals were used to flavour the wine-goblet.

This pretty specimen with fringed petals I brought from Switzerland. It is deliciously fragrant (D. barbátus). It grows in Roslin woods and on the banks of the water of Leith above Edinburgh. It is fitly named the Fringed Pink.

The Maiden Pink (D. deltoídes), with its small flowers and dark circle in the centre, and forked stem, is a trophy from Blackford Hill, near Edinburgh, where it flowers from July to September.

I have the Mountain or Cheddar Pink (D. cǽsius, *Plate IV.*, *fig.* 5), gathered and sent me from the Cheddar rocks in Somersetshire, so famous for their beautiful caverns. Its colour is fuller than that of the other Pinks, with the exception of the last-named.

Another family of the Pink tribe is that of the Catchflies. The puffed calyx of the Bladder Catchfly causes it to be so named (Siléne infláta). It grows on waste ground in all counties. The Latin name of the genus, Silene, was that of a youth in classic fable, who was employed by Minerva to catch flies for her owls to eat. One day she caught her messenger indulging in a nap when he should have been taking his prey, and the goddess, in angry indignation, turned him into a plant, and gave him a viscid juice upon his stem which should catch flies for ever.

The Sea Catchfly (S. marítima), closely resembles the Bladder Catchfly; but its calyx is not quite so inflated, and its flowers are larger and of a clearer white. Altogether, it is a prettier plant. Fanny describes it as mingling with the bright blue Bugloss and yellow Charlock in the gay drapery of the cliffs at Looe, in Cornwall; and it has been sent to me from various places on the south coast.

The English Catchfly (S. ánglica), is a much less graceful plant. The flowers grow from the sides of the scarcely-branched stem: they are pale pink. Fanny gathered this specimen in a corn field close to Marazion, near Penzance. The whole plant is very hairy and sticky.

The Variegated Catchfly (S. quinquevúlnera), is very pretty, the white petals being adorned with crimson blotches. The stems are forked and the leaves small. It grows on waste ground about Clevedon, where Fanny got her specimen.

The Nottingham Catchfly (S. nútans), also favours the coast of Somerset. It is an unassuming flower with deeply-indented white petals. It is very fragrant at night, and droops gracefully from rocks.

The Night-flowering Catchfly was found between Thirsk and Ripon, in Yorks. It is rare, has pale flowers opening at sunset, and closing by day, and its stem is much forked.

The rare species, erroneously named "Common" Catchfly, and the Striated, and Italian species, are very doubtful natives.

I first saw the beautiful little Moss Catchfly (S. acáulis), in Switzerland, whence my brother-in-law brought me a sod of it, as large, and just fitting into, the crown of his straw hat. The pot was covered with the large rose-coloured blossoms, so that only here and there a portion of the moss-like foliage was seen. A similar but smaller piece was afterwards brought to me from Snowdon. The stem is only half an inch high, and bears the upright bell-shaped blossom on its summit.

Next to the Catchfly family comes that of the Soapwort, of which we have but one British species (Saponária officinális, *Plate IV., fig.* 10). It is the handsomest plant of the tribe, grows two feet high, and bears a cluster of from five to ten pale pink flowers at the summit of its wand-like stem. It grows freely on the Castle Hill at Richmond and near Catterick Bridge. It is more plentiful among the Alps than with us; and, on account of its soapy quality, the Alpine Shepherds use it to wash their flocks. A decoction of the plant, boiled, makes a sufficiently good lather to wash linen.

The next family of the tribe is the Campion family, which

contains two or three familiar species. The Ragged Robin (Lýchnis flos-cúculi), with its bright rose-coloured torn petals, has always been a favourite with me. The poet Twamly was evidently very fond of the flower, and resents its being called "Ragged."

> "A man of taste is Robinet,
>   A dandy spruce and trim;
> Whoe'er would dainty fashions set
>   Should go and look at him.
>
> "Rob scorns to wear his crimson coat
>   As common people do;
> He folds and fits it in and out,
>   And does it bravely too.
>
> "Oh! Robin loves to prank him rare
>   With fringe, and flowers, and all;
> Till you'd take him for a lady fair
>   Just going to a ball.
>
> "Robin's a rogueish merry lad,
>   He dances in the breeze,
> And looks up with a greeting glad
>   To the rustling hedgerow trees."

There is enough of viscid juice about its joints to qualify it for a place among the aristocratic Flycatchers. Edward gathered some of these specimens near Hawkhurst, in Kent, and I have brought some of them from moist Yorkshire meadows. Hooker says that the name Lychnis is derived from the Greek word for lamp, because a cottony substance upon the leaves of an allied plant used to be made into lamp-wicks. Flos-cuculi means Flower of the Cuckoo; I suppose the season of the Cuckoo and of the flower being the same accounts for the application in this instance.

The Red Campion (Lýchnis dioíca, *Plate IV.*, *fig.* 6), is a frequent ornament of hedgerows; its bright petals are notched in the centre, the whole plant is hairy; its principal peculiarity is that it bears its stamens upon one plant and its stigma upon another. The reddish-rose colour of its blossoms makes it a very lovely ornament of the green hedge, affording one of the most pleasing contrasts imaginable. Leigh Hunt in his "Song of the Flowers" celebrates these colours, as well as praising other favourite flowers treated of in this chapter.

> " See, and scorn all duller
>   Taste, how Heaven loves colour ;
> How great Nature clearly joys in red and green ;
>   What sweet thoughts she thinks
>   Of Violets and Pinks,
> And a thousand flushing hues made solely to be seen.
>   See her whitest Lilies
>   Chill the silver showers,
> And what a red mouth has her Rose, the woman of the flowers.

> " Who shall say that flowers
>   Dress not heaven's own bowers?
> Who its love without them can fancy, or sweet floor ?
>   Who shall even dare
>   Say we sprang not there,
> And came not down that love might bring one piece of heaven the more?
>   Oh! pray believe that angels
>   From those blue dominions,
> Brought us in their white laps down, 'twixt their golden pinions."

There is a White Campion (Lýchnis vespertína), which grows in corn fields, and becomes very fragrant in an evening. It greatly resembles the Red Campion, so much so as to be considered by some high authorities a mere variety. I have gathered it near Clotherholme farm, a couple of miles from Ripon, and Edward has specimens from Kent.

The Viscid and Alpine Campions are mountain plants, and we have not obtained specimens of either.

The Corn Cockle represents the last family in the Pink tribe. Its botanic name (Agrostémma githágo, *Plate IV.*, *fig* 7), signifies "crown of the field," and suits the bright stately plant exceedingly well. The calyx indicates a state of transition between the two tribes, being united in the lower part like the Pinks, &c., and expanding in the upper into free leaflets in the style of the Chickweeds. It adorns corn fields in most of the counties of England, and its full purple colour and tall upright stems make it generally a favourite.

We now come to an extensive tribe of very humble plants which possess little interest except for botanists. The Pearlwort is the first family in the Chickweed tribe, and it has four British members. The Procumbent Pearlwort (Sagína procúmbens), is a mossy-looking plant, which spreads its small branches over neglected garden walks, or disused city paths, mingles with the grass of the lawns, or insinuates itself into the pots in a greenhouse. A cursory observer would say it had no flower; yet if you look close you will find small green crosses terminating the branches during the whole summer. These crosses are formed of the four sepals, which remain until the fruit is ripe; the petals are white, but much smaller than the sepals, and very generally absent altogether.

The Hairy Pearlwort (S. apétala), is still more frequently petalless; indeed, the absence of petals and the hairiness are its distinguishing characteristics. Fanny brought these specimens from a wall on the margin of Fowey harbour, and she found the Sea Pearlwort (S. marítima), on the Looe cliffs. This species differs more from the Procumbent and Hairy ones than they do from each other; for its small leaves are seated on the stem at more distant intervals, thus taking off the mossy appearance; and its flowers are not so minute.

The Mœnchia is a nearly allied plant, but with bright starry flowers and spare foliage. It has also four sepals, four petals, and four stamens. Fanny found it spangling the turf on the moor above West Looe.

The Umbelliferous Chickweed (Holósteum umbellátum), has its pinkish flowers in a cluster. It is a rare plant, occasionally found about ancient ruins.

The Spurreys have five petals and sepals, and ten stamens. The Awl-leaved Spurrey is a slender, half-procumbent plant, with a great resemblance to the Sea Pearlwort. These specimens came from Fowey.

The Knotted Spurrey (Spérgula nodósa), has erect branches, and brilliant white flowers. It is a great ornament of marshy places on our Yorkshire moors.

The Pearlwort Spurrey frequents the Scotch mountains. We have no specimens.

At once the prettiest and most troublesome of the family is the Corn Spurrey (S. arvénsis). Its flowers are in a forked cluster, with frills of narrow leaves round the stem at every joint. Hooker says that it is cultivated and much esteemed in Holland. But on this point Dutch and English taste seem at variance; for here it is considered a troublesome weed, and in Norfolk it has the name of "Pickpocket."

And now we come to the true Chickweeds, the darlings of birds, though the despised of man. But we may claim interest even for the common Chickweed (Stellária média), because it was in observing this plant that Linnæus detected what he called the "sleep of plants." The pairs of leaves contract towards evening, and enfold the buds; they do the same before rain, and, if it be heavy, it is long ere they open again. It must be of such small flowers as these that Bulwer says—

> "Wearied children on earth's gentle breast,
> In every nook the little field flowers rest."

The clear white flowers of the Greater Stitchwort (S. holóstea, *Plate IV.*, *fig.* 8), are a familiar ornament of the hedgerows in June and July, inviting the eager hands of flower-loving children, and disappointing them as surely by its speedy fading.

The Smaller Stichwort (S. gramínea), is also common. Its stem is not rough, like that of the Greater one, but smooth and bright green, and the leaves are grass-shaped. Its slender white petals, cloven to the base, are relieved by crimson anthers. It grows among bushes in damp places in hilly districts.

The Wood Stitchwort (S. némorum), I gathered on Brignall banks, near Rokeby. The steep wood was gay with the thousand white stars produced by this plant, which gleamed from a background of their own abundant pale green foliage. Here the leaves are broad and large, forming a great contrast to the last species.

The Bog Stitchwort (S. uliginósa), has the smallest flower of all; its sepals are pointed, and twice the length of its tiny petals; the flower-branches grow from the juncture or *axils* of the leaves, and the foliage is of a glaucous hue. It flourishes in wet places, and we have plenty of specimens both from Kent and Yorkshire.

We have none of us found the Marsh or Alpine Stitchworts.

The second family in the Chickweed tribe is that of the Sandworts. The handsomest species of these is the Sea Sandwort (Arenária marítima), which Fanny found on the shore at Looe. Its lilac flowers are bright, and have a yellow blotch in the centre.

> "Among the loose and arid sands
> The humble Arenaria creeps;
> Slowly the purple star expands
> But soon within the calyx sleeps."

The little Purple Sandwort (A. rúbra), is an inhabitant of sandy lanes. It is a miniature of the Sea Sandwort. I found it in Cheshire some years ago, and Edward has specimens from the sandy lanes of Kent. Both these species have awl-shaped leaves; but those of the Sea Sandwort are blunt, and those of the Purple Sandwort end in a bristle point.

On the shore near Penzance Fanny found abundance of the common Sea Sandwort (A. peploídes), with its small white flowers, fleshy leaves arranged in two rows, heavy stems, and large seed-vessels.

The Plantain-leaved Sandwort (A. trinérvis), grows plentifully in Swaledale, upon the margin of mountain streamlets; its leaves are very large with three marked ribs, and its flowers very small.

The Fine-leaved Sandwort (A. tenuifólia), I got off the wall of Easby Church, it grew also on the ruins of the fine old Abbey; the plant is slender and bushy, and soon withers away.

The Thyme-leaved Sandwort (A. serpyllifólia), I first recognised whilst looking from the window of a railway carriage as the tickets were being collected previously to the train entering the station. I was not able to go back when I reached the station, as a carriage was waiting to convey me far into the country; but I lost no time in writing to a friend in Richmond to procure the plant for me. She was much amused at being requested to botanise in so novel a field of research, but she quickly procured the plant; the flower is small, but the plant is compact, and the broad-shouldered leaves are not out of proportion.

The Spring Sandwort (A. vérna), is my favourite; its stems are light, and its leaves narrow, and it is half covered with brilliant white starry flowers. It is almost the only plant that is not injured by lead in the earth and water; but it grows as freely upon the heaps of rubbish near the mines, and upon the bed of the lead-stained stream, as upon the green margin of the most limpid brooks.

The Fringed and Level-topped Sandworts are mountain plants.

The Mossy Cyphel we have none of us found, I believe it to be very rare.

The Mouse-ear Chickweeds are the last family in this tribe; they vary from the Sandworts in having notched or cloven petals, while those of the Sandworts are entire.

The Water Mouse-ear is very like the Wood Stitchwort, but it has five stigmas, while the Stitchworts have but three.

The Field Mouse-ear (Cerástium arvénse, *Plate IV., fig. 9*), is the handsomest of the family; it has large intensely-white flowers, and its foliage is small and narrow, and of a white powdery hue. It is pretty as a rock plant, or for a border for Scarlet Geraniums. These specimens were from a gay plot on Richmond Castle Hill.

The Broad-leaved Mouse-ear (C. vulgátum and viscósum), has insignificant flowers and coarse rough foliage; it is unobjectionable when mingling with the sward, but is often a troublesome garden weed, as is also the Narrow-leaved species, the narrow foliage of which makes it more graceful-looking.

The Little Mouse-ear (C. semidecándrum), is stiff in its growth like a tiny shrub, and its petals are less cloven. I found it on the railway line in a similar situation to that of the Thyme-leaved Sandwort.

The Four-cleft Mouse-ear (C. tetrándrum), with its four petals and four stamens, forming a single exception to the five petals and ten stamens of the family, grows on Braid Hill, near Edinburgh.

The Alpine Mouse-ear we are not in possession of.

CHAPTER V.

TILIÁCEÆ — HYPERICÁCEÆ — MALVÁCEÆ — LINÁCEÆ — ACERÁCEÆ — HIPPOCASTANÁCEÆ — GERANIÁCEÆ — BALSAMINÁCEÆ—OXALIDÁCEÆ.

> " The leaf-tongues of the forest, the flower-lips of the sod,
> The happy birds that hymn their rapture in the ear of God,
> The summer wind that bringeth music over land and sea,
> Have each a voice that singeth this sweet song of songs to me:—
> This world is full of beauty like other worlds above,
> And, if we did our duty, it might be full of love! "
>
> GERALD MASSEY.

THE LINDEN order comes next of British groups in the natural arrangement; and although we have but few species, it is not wanting in interest. The characteristics are four or five sepals soon falling off, and the same number of petals, and numerous stamens. They resemble the Mallows in the valve-like arrangement of the calyx, but are distinguished from them by their free stamens.

The elegant and fragrant blooms of the Lime tree (Tilia europǽa, *Plate V., fig.* 9), attract crowds of bees, which keep up a perpetual buzzing among the leaves. I remember a swarm issuing from one of our hives, and making its way, as if by a preconcerted plan, to a group of Limes in the next garden. The foliage of this tree is very pleasing, appearing early and gladdening the eye all the summer with its fresh green hue. The leaves are heart-shaped, and a tuft of minute hairs is situated at the juncture of the veins on the under side. Several blossoms grow in a cluster, and a sort of thin leaf accompanies the flower-stalk part of the way. The Swiss planted these trees upon every victorious battle-field, and with them the Linden is regarded as an emblem of liberty. In Russia they use the inner bark of the Linden to make bast mats and ropes, which are the best suited for wells, as they do not rot with wet. The wood is applied to turnery purposes and for making vineyard ladders. So great is the value there attached to this tree, that the people are required by law to plant it on the roadsides. Linnæus took his name from this tree, which in the Swedish language is called Lin. A very large Linden growing near his father's house in his native village suggested to the philosopher the idea of assuming that as his name.

Another species of the same genus furnishes the Indians with fishing-tackle.

We have a Downy Linden and a Small-flowered Linden as well as the common Lime. Edward brought specimens of the two latter trees from the neighbourhood of Hawkhurst, in Kent; they have smaller darker leaves, and the Downy species is more hairy.

Jute hemp is the produce of a plant of this order, a native of Bengal. It grows to the height of twelve feet, and its bark contains valuable fibre. The hemp formed from this fibre is fine and satiny, and is used in India for making bags and wrappers, and in England for mixing with silk in the manufacture of cheap satins.

The ST. JOHN's WORT is a small order, chiefly interesting as including the valuable plant or shrub which produces the tea of commerce. This commodity, though only introduced into Britain less than two hundred years ago, has become to us a necessary of life. China is the natural home of the tea plant, but one species is successfully cultivated in India. The black and the green tea are leaves off the same tree, but in different stages, and prepared in a somewhat different manner. The characteristics of this order are five sepals and five petals, and numerous stamens united into several groups: on this account the order is designated "many brotherhoods" in the Linnæan system. The fruit is a berry with several cells. The order includes trees, shrubs, and herbs. The blossom of the Large-flowered St. John's Wort would satisfy the most gaudy taste, contrasting its ample yellow petals and tassels of yellow stamens with the massive dark green leaves. When I was a child we use to call this flower the Rose of Sharon; and I reverenced it for its scriptural mention. Now that I know that a very different flower is the one used as an emblem of our Saviour I still love the gaudy St. John's Wort for the sake of old association. It grows wild in some parts of Scotland and Ireland, but my specimen is a shrubbery one (Hypéricum calycínum).

The Tutsan (H. androsǽmum), or Park-leaves, has foliage resembling that of the large-flowered species; it is a larger and more shrubby plant, and its flowers grow in small clusters of three or four. The calices are often crimson, and the young buds have a waxy look like yellow berries within their cups. The flower is small compared with the last species, but the large black berry attracts much attention. Edward found it in the lanes about Hawkhurst last July, and I saw the black berries and richly-tinted autumnal leaves about Clevedon late in the year.

The Kentish lanes also furnish abundance of the square-stalked St. John's Wort or St. Peter's Wort (H. quadrángulare); it has pale flowers small and very plentiful, and abundance of little dotted leaves. It flowers early in August.

The hedge banks in the same neighbourhood were adorned six weeks earlier with the clear yellow stars of the elegant little trailing St. John's Wort (H. humifúsum), a pretty prostrate cheerful plant.

The Perforated St. John's Wort (H. perforátum), is so called because of the numerous transparent glands scattered over its leaves, which give them the appearance of being perforated. Several of the other species have these glands also, but none in so great a degree; its petals are dotted and striped with black.

The Upright St. John's Wort (H. púlchrum), is a pretty species, of slender growth, and smaller clusters of flowers; the buds and young shoots are tipped with crimson.

The Hairy St. John's Wort (H. hirsútum), is distinguished by its woolly stem; it, and the two preceding it, are all very common, flourishing in woods and field borders in Yorkshire, Kent, Somersetshire, &c. (*Plate V., fig.* 6). I fancy these species were confounded by the old herbalists, and that the mystic virtues of the one were extended to all. St. John's Wort used to be employed to expel demons. People dressed their houses with it on St. John's-eve, and in France and Germany branches of it were placed at every door on that occasion. The yellow juice of this plant closely resembles the gamboge of commerce. The Perforated species was formerly called the " herb of war," its perforations were supposed to resemble wounds ; and, consequently, by the curious reasoning of analogy, which was then in vogue, it was argued that it would heal them.

" Hypericum was there, the herb of war,
Pierced through with wounds, and marked with many a scar."

The Marsh St. John's Wort (H. elódes), I found in a peat-pool on Rudd Heath, in Cheshire, a famous field for botanic research; but which cultivation is fast "reclaiming," as it has done many a rich bog and fen. It is a thick succulent plant, and, though growing an inch deep in water, was still clothed with dense wool. The plant is only a span high, and its flower is very small.

Fanny has a Cornish member of this family, the Mountain St. John's Wort (H. montánum); its flower is as large as that of the Perforated species, but its colour is pale like St. Peter's Wort; it grows in hilly groves about Looe.

The MALLOW order succeeds that of the St. John's Wort. Here the sepals and petals are five, the latter twisted when in bud. These are hairy plants or shrubs; their qualities are harmless, and in some instances useful. The rich-coloured Hibiscus, and the stately Hollyhock belong to this group, and there are foreign species still more beautiful. The Cotton plant belongs to this order, and it is indigenous in Asia and America. Many species are now cultivated in extensive districts between the tropics. Cotton is a woolly substance which envelopes the seed; it has been used in Egypt from times of great antiquity, and several remains of Greek literature have been found written on cotton. It is now an article of commerce all over the world, especially important in the British trade as giving employment to thousands of labourers of both sexes and all ages. The true Mallows have an inner and outer calyx; the former comprised of five valves, the latter of three sepals united at the base. The name is from a Greek word signifying *soft*, and refers to the emollient nature of the plants.

The common Mallow (Málva sylvéstris), grows freely on waste places. It is a strong woolly plant, with large lobed leaves and crimson flowers, handsomely marked with a darker shade. It grows much in cottage gardens in the north of Yorkshire, being cultivated and used for fomentations in case of swellings, toothache, &c. All children love to play with the miniature "Cheeses" which form the seed, as Clare says—

> "The sitting down when school was o'er
> Upon the threshold of the door;
> Picking from Mallows, sport to please,
> The crumbled seed we call a Cheese."

French children play the same game, terming the seeds "les petits fromageons." Syrups and pastils are made from this plant. In Job's time it was used as an article of food, "Who cut up Mallows by the bushes, and Juniper roots for their meat" (Job xxx. 4); and Dryden speaks of a similar custom—

> "Shards and Mallows for the pot."

The Romans used to cook and eat it as a vegetable, and it is still cultivated in Egypt as an article of diet.

The Round-leaved Mallow (M. rotundifólia), is a smaller plant with pale flowers. It has the same qualities as the common Mallow; but being smaller and less frequently met with, it has been little noticed.

The Musk Mallow (M. mosc háta), is the handsomest member of the family. Its petals are of a delicate pink, beautifully veined with rose. Like its two companions, its stamens are united into a single "brotherhood," and its leaves are musk-scented, and divided into deeply cut lobes. The common Mallow begins to flower in June, and the other two in July; they flower through August and September.

A near relation of this is the Tree Mallow (Lavatéra arbórea, *Plate V., fig.* 2), the most stately-flowering in our Flora. Fanny treasures it in memory of a pleasant trip to Looe Island, off the coast of Cornwall, near the two small towns from whence it takes its name. The channel between the island and mainland is narrow and shallow; and is rendered very dangerous for vessels, except at high water, by a rock which rises midway called "Brown Meg." A schooner was stranded there when Fanny and her friends crossed; and though she was afterwards set afloat again, yet the loss by injury to her cargo, &c., was very serious. They wandered over Looe Island, from its commanding peak to its rocky shore, where the crevices were crowned with the proud form of the Tree Mallow. The leaves of this plant are dull green, and of soft velvety texture, and its purple flowers are shaded almost to black in the centre, while the brotherhood of light stamens stands prominently out from the deep dark cup. Some of the plants were six or eight feet high.

The Marsh Mallow (Althǽa officinális, *Plate V., fig.* 2), grows in swampy ground on the coast of Kent; its foliage is of the same dull hue as that of the Tree Mallow, and its tall upright stems are clothed with velvety hair. Its blossoms are clustered and of a very pale pink, and the stamens are crimson. This plant abounds more in the softening mucilage so valuable in medicine than any other of the order; it is sold in a dried state, and an infusion made from it for rheumatic and neuralgic pains, and for sore throats; and from the juice cough lozenges are made.

In passing from the Mallow to the FLAX order we exchange cotton for linen, and lose sight of the coloured half-clothed cultivators in contemplating European Flax fields. In this

F

order the flowers have five sepals and five petals, which, like the Mallows, are twisted when in bud, they are very perishable, falling off with the slightest touch or puff of wind; five stamens united at the top of the ovary, and five stigmas.

I have specimens of the common Flax (Línum usitatíssimum, *Plate V.*, *fig.* 1), from a field near Warminster, in Wilts; and I have occasionally seen a plot of it about Richmond, cultivated I should imagine for its seeds. The petals are notched, and of a lovely blue. The Latin name, signifying "most useful," is justly applied to this plant; for the firm fibres of its stem supply us with linen, the oil expressed from its seed is valuable for mixing paint and other purposes, and meal formed of the ground seed makes excellent poultices, and is also used as a medicine. It is cultivated in Devonshire, where Carrington describes it—

"How sweetly blows
Upon the slopes the azure-blossomed Flax!"

The Narrow-leaved Flax is more common as a wild plant. Edward has brought specimens from Kent, Fanny from Somersetshire, and I from Yorkshire. The principal distinction between the species is the notch in the petals, which is present in the common one and absent in this. The Narrow-leaved Flax is even found in the Highlands of Scotland, and receives there the name of the Lintbell. The simple and poetic inhabitants of these lovely districts are close observers of nature, and they not unfrequently reply to your questions in figurative language worthy of the flowery parlance of the East. Thus, a poor Highland woman, unable to read or write, could yet note how the frail petals of the Flax unfolded in the sunshine and closed when his rays were withdrawn; she was ignorant of everything, her one power being to accept and love the blessings of God, from the "inestimable gift" of His Son to that of the least flower that bloomed in her path. Being questioned as to her adherence to the profound doctrines of the Kirk, she showed such utter ignorance that the minister deemed her totally unfit to become a communicant. He conveyed this to her as kindly as he could, and she replied, "Aweel, Sir, aweel; but I ken ae thing—as the Lintbell opens to the sun, so does my heart to the Lord Jesus!" Here was seed of the Lord's own planting; no hireling had cultivated this ground, or scattered into it the germs of life, but the Second Adam had been busy in this garden, dressing it and keeping it.

The Perennial Flax is the handsomest and most scarce member of the family. It was gathered on Leyburn Shawl, an extensive table-land overlooking the rich valley of Wensleydale, in Yorkshire. It commands a view of Bolton Castle, where the unfortunate Mary Stuart was imprisoned. She succeeded once in effecting an escape, but her flight was soon discovered, and she was recaptured on Leyburn Shawl. This plant, with its lovely but evanescent flowers, is a fit memento of that beautiful and frail woman.

The Small Cathartic Flax (Linum cathárticum), is very different from its brethren; its petals are snow-white, its leaves grow opposite one another, and the flowers droop. It often mingles with the herbage in meadows, but, having biting qualities, I dare say the cattle would as soon be without it.

The Little Flax-seed (Radiola millegrána), is the smallest of our flowering plants. Fanny found a number of the plants on the right bank of the Loe Pool, a small lake near Helstone, in Cornwall, only separated from the sea by a bar of sand. There is an interesting description of it in Mr. Johns' "Week at the Lizard," which you must read some evening when you have leisure. Fanny determined to visit the Loe Pool because of the plants which he described as growing around it, and she persuaded her mother to stay at the snug "Angel" inn, at Helstone, for that purpose. The number of plants she got fully rewarded her for all her efforts, and this tiny Flax-seed was one. The tiny flowers have four sepals, four greenish-white petals, four stamens, and two stigmas.

We now come to another family of trees, the members of which are familiar to us in every stage of growth. In the winged seed of the SYCAMORE may be found a miniature representation of the young plant, with its stems and leaves; near the parent tree we soon see a pair of leaflets spring, and in a few months a fresh shoot rises, and an exquisitely-folded palmate leaf begins to open. A little higher and a little higher marks the progress from year to year, until the stages of childhood and youth are past, and the fine forest tree has attained maturity, and stands in the full dignity of blossoms and fruit.

> " Then he spoke to those wood-dwellers,
>   ' Ye are like to men,
> And I learn a lesson from ye
> With my spirit's ken :
> Like to us in low beginnings,
> Children of the patient earth,
> Born like us to rise on high,
> Ever nearer to the sky,
> And like us by slow advances from the minute of your birth.' "
>
>   CHAMBERS' *Journal*.

The Latin name for this family, *Acer*, means sharp, because in olden times sharp instruments of war were made of Sycamore wood. In the present day it is principally employed in making platters, bowls, and musical instruments. This tree lives to a great age. There is a Sycamore in the county of Edinburgh known to have been planted there before the Reformation. The Egyptians held the Sycamore in high estimation, and in them they were chastised. "He smote their Vines with hail, and their Sycamore trees with frost." The trees of this family have blossoms with five sepals, five petals, and eight stamens.

The Maple is a low-growing tree, seldom rising much above the hedgerow; its leaves are smaller than those of the Sycamore, and its flower-cluster is erect, not pendant as in that species. In both the winged seeds hang in graceful groups, and receive tints of crimson from the summer sun. The bark is very deeply furrowed, and the Kentish farmers value it on this account for Hop-poles, the rough surface affording warmth to the young Hop vines. The wood of the Maple is very beautiful when polished. Pliny relates that the luxurious Romans would give immense sums for tables made of this wood, and that when these gay lords reproached their ladies with the extravagance of their dress and jewellery, the fair ones retorted by inquiring the price of the Maple tables. There is a somewhat sugary quality in our Maple (Ácer campéstre), but it exists in a very slight degree. But in America there is a species called the Sugar Maple, which notable housekeepers tap annually, and from the sap manufacture sugar enough for their domestic needs. The seed-lobes of another species are cooked by the Tartars as an article of food, but it is no great compliment to any seed to be eaten in a district where vegetation is so scanty.

The HORSE CHESTNUT family is nearly allied to that of the Maples, and should surely now be admitted among British trees. It is a native of Asia, and is spoken of by Gerarde as a rare foreign tree. Mr. Johns states that in some places it is called "the Giant's Nosegay;" and when I have looked upon a Horse Chestnut covered with its magnificent clusters of pink and white flowers, I have been struck with the suitability of this popular name. The fruit is too bitter to be pleasant for food, but the Swiss make good use of it for fattening cattle.

The GERANIUM or CRANESBILL order is the next in our British arrangement. These plants have five sepals, five petals, ovary in five divisions, five styles, and ten stamens in the brotherhood. Our wild Geraniums are all mere herbaceous plants; but the Pelargoniums, as the greenhouse Geraniums are now called, attain to the dignity of shrubs. The long beak to the seed procures for them the English name of Cranesbills.

The Bloody Cranesbill (Geránium sanguíneum), is a rare and handsome species, with a deep crimson flower, and half-prostrate stem. I have never found it but in Switzerland, where I gathered it from rocks overhanging the Lake of Thun; but it grows freely near Whitty. Fanny had the pleasure of finding it in Cornwall. She was driving from Helston to the Lizard Point with her mother, and, as they knew the tide was low, they stopped at Kynance, sent the carriage on, and proceeded on foot along the cliffs. Kynance Cove was looking extremely lovely; its distorted rocks of serpentine, deeply tinged with crimson, maroon, and green, stood out boldly from the pearly sands. The rich dark tints of these rocks she describes as giving a very peculiar character to the landscape, while they furnish a congenial habitat to many plants rare elsewhere. The Bloody Cranesbill is one of these, and grows in luxuriance on the cliffs for some distance beyond Kynance, and the English Scurvy Grass grew in damp places in the same vicinity.

The Least Cranesbill (G. pusíllum), Fanny got near the Land's End, at the edge of the road leading to the hotel. It is like a miniature of the Dove's-foot Cranesbill.

The Dove's-foot (G. mólle), has round leaves, rather notched than lobed and very downy, and small purple flowers. The whole plant has a musky odour; it is very common on the margin of fields and lanes.

The Herb Robert (G. Robertiánum), is always welcome, appearing, as it does, among our early spring flowers, and plentifully adorning the hedgebank with its rose-coloured flowers and crimson-tinted leaves during most part of the summer. Its strong aromatic perfume is generally considered an additional recommendation.

The Long-stalked Cranesbill (G. columbínum), I found in Wiltshire, in a lane between Longbridge Deverill and Horningsham; it is taller, its lilac flowers are larger, and it is a more slender plant than the rest. But its greatest peculiarity is an elastic power in the seed-pod, which enables it to throw the seed out with force. This elastic carpel contracts in dry weather, and expands in wet; the seed is very beautiful, as well as very curious, each carpel bends outwards, and contracts again above—they thus form five open loops. I found both flowers and fruit upon the plant in August.

The Jagged-leaved Cranesbill (G. disséctum), grows very luxuriantly in stubble fields in the same neighbourhood. It is a common wayside plant; its leaves cut nearly to the middle, and its purple flowers, distinguish it from the other small species.

The Larger Dove's-foot (G. pyrenáicum), is a handsome plant, the bloom twice the size of the Slender-stalked species, and the lobed leaves are kidney-shaped. My specimen was sent me from the Isle of Wight, and at the same time I received a piece of the Round-leaved Cranesbill, the carpels of which are curved in the way I have before described.

The Shining Cranesbill (G. lucídum, *Plate V., fig.* 5), is a pretty ornament of stony places; its round, lobed leaves are thick and glossy, and very often edged with scarlet: its flowers are bright pink. It grows abundantly about Richmond, in Yorkshire.

The Dusky Cranesbill (G. phæum), is one of the most scarce of the family; it grows under Fremington Edge, near Reeth. The petals are a brown crimson shading almost to black, but white towards the point, so as to give the flower the appearance of having a white eye. It reaches the height of a foot, and is pretty often to be found in old-fashioned gardens.

The Wood Cranesbill (G. sylváticum), is a frequent ornament of the woods and hilly meadows in the neighbourhood of Richmond. Its flowers are large and of a crimson purple.

On leaving the hilly districts and descending to the flatter country, the Meadow Cranesbill (G. praténse, *Plate V., fig.* 4), with its large intensely blue flowers, takes the place of the more alpine species. It is surpassed by none of them in beauty, and is a very great favourite with all flower-lovers.

Only one species of this large family is now wanting, and that was supplied to me last year from the very clever and obliging Curator of the Botanic Garden at Edinburgh. It has a crimson flower not unlike that of the Wood Cranesbill, and a ternate or five-lobed leaf. It is called the Knotted Cranesbill (G. nodosum).

The Striped Cranesbill (G. striátum), grows in Aske woods, a part of Lord Zetland's grounds; it is a beautiful species, the white petals covered with purple veins. It has recently been admitted into our British Flora.

The Storksbills are closely allied with the Cranesbills, but have a longer beak and pinnate leaves.

The Hemlock Storksbill (Eródium cicutárium, *Plate V.*, *fig.* 8), is a common plant about Edinburgh, adorning the rocks on Braid Hill and Arthur's Seat, and the shore along the Firth of Forth. Its bright pink flowers are very pretty.

The Sea Storksbill (E. marítimum), has no beauty. Its tiny flowers are generally petalless, and it would attract no attention except from a botanist. Fanny got it both at Looe and Polperro, in Cornwall. At first she fancied that the petals had fallen off by accident, and she kept the plant in water until a fresh bud should open; this was also petalless. She tried it again, but with the same result, and afterwards she found from some book on botany that the petals were generally absent.

All the members of this family have a musk odour, but the Musky Storksbill (E. moschátum), has it in much the greatest degree. This plant inhabits mountain pastures; it differs but little from the Hemlock Storksbill, and many eminent botanists think it merely a variety of that species.

The Garden Rues, much recommended as a "Bitter," and the Fraxinella, belong to orders succeeding that of the Geranium, though without British representatives. The Fraxinella is a native of Switzerland, and as a handsome shrubby plant has gained entrance to our gardens. The whole plant has a strong odour, agreeable to some, but very much the contrary to others. Dr. Murray, in his book on "Vegetable Physiology," states that the atmosphere around this plant becomes luminous in hot weather.

The TROPÆOLUM family, so well known in our gardens by its representatives the common Nasturtium and the plant known as "Canariense," is nearly allied to the Geraniums, being chiefly distinguished by the spur at the back of the calyx. The half-ripe fruit boiled in vinegar forms an excellent substitute for capers.

The BALSAM family has British representatives. The Touch-me-not (Impátiens noli-me-tángere), was found wild in the neighbourhood of the English Lakes. I had before dried specimens which I found in Switzerland, at the foot of the Staubbach waterfall. I have also seen it in shrubbery gardens in Wiltshire, where it is self-sown, and in Mr. Ward's botanic garden at Richmond. The whole plant is very succulent; the corolla is of an irregular form prolonged into a spur, altogether shaped somewhat like a cornucopia; it is yellow, dotted with red. The reason of its name, Touch-me-not, lies in the five-valved seed-vessels, which burst with the slightest touch.

The Tawny Touch-me-not differs little except in the red tinge of the corolla from the common species. It is only found in Surrey. My specimen was the gift of Mr. Ward. The handsome double Balsam cultivated in our greenhouses is an Indian brother of the Touch-me-not.

The OXALIS or WOOD SORREL (Óxalis acetosélla, *Plate V., fig. 7*), comes next in rotation. There is only one family in the order. The common Wood Sorrel has delicate triple leaves, often lined with purple, and it flourishes under the deepest shade. Its flower has five sepals, five petals, ten stamens, and five styles; it is generally drooping, and its pale pencilled blossoms have a most pleasing and modest appearance. Old Gerarde calls it "Alleluya" or "Cuckoo Meate," because, he says, "either the cuckoo feedeth thereon, or by reason when it springeth foorth and flowreth the alleluya is sung in the churches." The seed-vessels when ripe bend downwards, and are hidden by the leaves: hence the plant was supposed to have no seed. The whole plant has a sour flavour, and from its leaves the poison called oxalic acid is obtained.

There is a yellow Oxalis with a taller stem and small flowers growing in clusters of two or three. It grows wild in neglected gardens, and a few warm woods. My specimens were from a garden at Brixton, in Wiltshire; none knew how it came there, but each year fresh plants appeared among the Mint. Mr. Ward had it in his botanic garden (O. corniculáta).

This order concludes the Receptacle subclass, the grand distinctive feature of which is that all the parts are built on the receptacle. One occasional supplementary part of a plant we have left unnoticed, though it exists in several of the plants we have handled. In the Sea and Purple Sandworts, each little cluster of leaves has a large semi-transparent scale which covers its juncture with the stem, this is called a *stipule*. These leaf-like appendages vary from the simple form of a mere scale to the full structure of a real leaf. In the Lime the *bract*, for so it is called when accompanying the flower-stalk, is paler than a leaf, though veined like one; in the Willow-herb family it is the same colour as the leaves. The Samphire has a cluster of flowers (*Plate IX.*, *fig.* 7), which is called an *umbel*. Where the flowers join the main stem there is a row of *bracts* all round; this collection of bracts is called an *involucre* or wrapper. Compound flowers, as the Thistle (*Plate XI.*, *fig.* 3), have row upon row of these bracts, or what is called a *compound involucre*. The acorn-cup is another form of united bracts. When such leaflets occur at the foot of the leafstalk, instead of the flower-stalk, as is the case with the Rose-leaf and the Sandwort, they are called stipules.

ROOTS.

The *root* is a very important part of the plant, and we ought not longer to delay a study of it. The so-called "creeping roots" of Solomon's Seal (*fig.* A), Anemone (*fig.* B), Coralwort (*fig.* C), and Water Lily, are merely underground stems with little roots at their joints. They are marked with scales, which appear to be the rudiments of imperfect leaves. The Creeping Crowfoot (*fig.* D), has prostrate stems rooting at the joints,

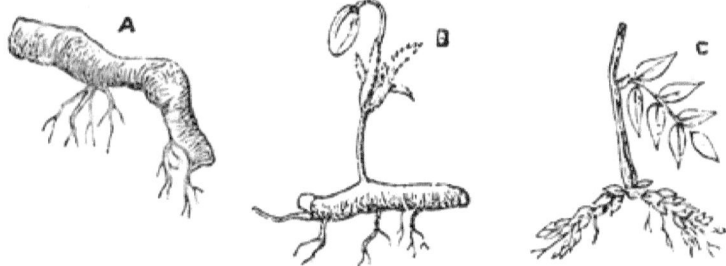

and the offshoots of the Couch-grass are but young underground stems. We always hear "bulbous roots" spoken of, and these again are not real roots, but contracted stems and leaf-buds. We call them *corms* (*fig.* E). If the corm of a Crocus be cut

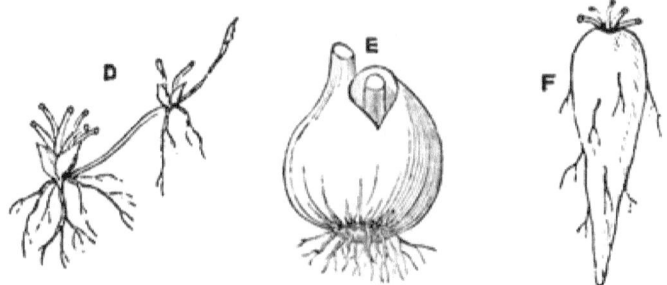

in two just before the leaves begin to appear, it will be found to consist of a *collar* from which the true roots spring, a solid white part which is the stem, and one or two little buds in the upper parts, which contain the germs of leaves and flowers. The old corm wastes away when it has done flowering, and a new corm is formed beside it. In the Crocus the new corm forms above the old one, and the remains of the old one hang like a fibrous ring at the base of the new.

The first action of a germinating seed is to send down a tap-root; as the plant increases in size this generally disappears, and the root assumes the characteristic form of the species. Herbaceous plants have generally either fibrous roots, the fibres extending perpendicularly, or horizontally, as the case may be; or thickened roots, in which a store of mucilage is laid up for the use of the plants. The *spindle-shaped* root (*fig.* F), is of this description, and many plants have it—for instance, the Carrot, Parsnip, Radish, &c. The *truncated* root of the Devil's-bit (*fig.* G), is also a storehouse for the plant's nourishment;

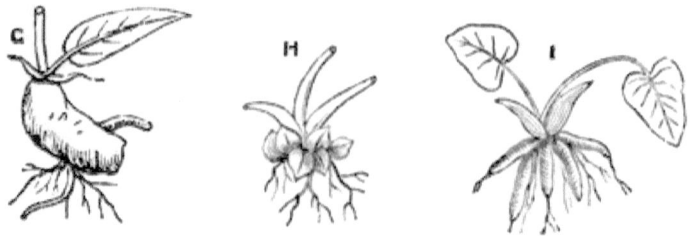

the granules of the Meadow Saxifrage (*fig.* H), serve the purpose, and the Lesser Celandine stores its mucilage in *bundle roots* (*fig.* I). Some botanists class the *tubers* of the Orchis

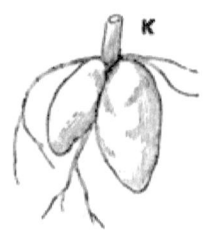

(*fig.* K), with these thickened feeding-organs, and others consider them more of the nature of underground stems.

But whether the granule, tuber, or spindle be present or not, the plant is supplied with *fibres*, simple or branched, perpendicular or horizontal. These are the mouths of the vegetable, by which it imbibes moisture and nourishment from the earth; the moisture rises through vessels which traverse the whole plant, be it herb, shrub, or tree, reaching to the leaves and the remotest bud before evincing any change; there it permeates the lungs of the plant, becomes converted into characteristic sap, and gradually descends again by a separate machinery of vessels. The structure of the woody parts of roots corresponds generally with that of the stem, only instead of pith it has in the centre a bundle of woody fibre and vascular tissue. The true root is divided from the stem by a *collar*, all the parts beneath the collar, whatever their form and nature, may be said to belong to the root, though they may not serve the purpose of the root. Possibly it may not have occurred to you of how great a service roots are. The utility of roots as the mouths of the plant I have already mentioned, and they are of equal importance as anchors to the plant. The roots of the Oak extend in every direction as far as the branches, and this accounts for the great stability of the tree and its security amid storms and floods. The fibrous roots creep always in the direction of water, they have enormous power in penetrating the smallest crevices in rocks, encircling stones, and pushing their way through every impediment, thus making good a strong position for the plant. Some roots are very aggressive in their habits. Horseradish and Mint are very difficult to restrict to the ground assigned to them; and the Couch-grass, against which the farmer preaches an everlasting crusade, finds means of overcoming every check to its progress. Dr. Murray found bulbs of the White Lily threaded by fibres of the Couch-grass like beads on a string. Yet even these obtrusive grass-weeds have their utility, binding securely the sand on the river-bank or sea-shore, and thus forming more enduring dykes than the hand of man could raise.

It sounds selfish to speak of the edible qualities of roots as their highest utility; but it is not really so, for did not God create the herb and grass of the field, and every tree, and give them to man with the clear injunction, "To you it shall be for meat?" In a spirit, therefore, of thankfulness to our beneficent Father, I would enumerate the articles of food which He has made to grow for us in the roots of plants—the Carrot, Turnip, Radish, Artichoke, Beetroot, Onion, Parsnip.

Instinct prompts the most savage nations to dig up and eat roots, and great numbers subsist upon such food, thus contenting themselves with the humblest of God's provisions. Alas! that the image of the Destroyer should so efface that of the Creator in the highest of His works, that we should hear of the fierce savage killing his own children, because the cravings of hunger prompted them to steal the roots he had dug up!

CHAPTER VI.

CELASTRÁCEÆ—RHAMNÁCEÆ—LEGUMINÓSÆ.

> "There is a lesson in each flower,
> A story in each stream and bower;
> On every herb on which you tread
> Are written words, which, rightly read,
> Will lead you from earth's fragrant sod
> To hope, and holiness, and God."

In the plants which we are now about to study we must no longer look for the stamens to be situated upon the receptacle—that was the character of the Thalamiflorals; the second subclass, that of the Calyciflorals, makes the calyx the bond of union, and the stamens are situated upon it.

The twenty-fourth British order, the first in this subclass, is that of the SPINDLE TREE. The only British representative of this family is Euónymus europǽus, a pretty shrub with lance-shaped leaves, smooth branches, and small clusters of greenish flowers. These have generally four petals, four sepals, and four stamens, though the number is sometimes five. The blossom is unattractive, but the seed is very beautiful, consisting of a fleshy crimson vessel, divided by a depressed line crossing it at right angles into four parts; the vessel opens at these lines, and discloses four large seeds of a brilliant orange. I have found this shrub about Richmond, in Yorkshire, and on the borders of the Downs in Wiltshire, and Edward has it from Kent. No great interest attaches to this shrub; butchers' skewers are made from it; but, what would be more likely to redeem it from neglect is, that it is good for making piano-keys. The Arabs eat the green leaves of one species of Spindle, and wear a sprig of it to preserve them from the plague.

The common BUCKTHORN (Rhámnus cathárticus), is found

in hedges and thickets on the borders of Salisbury Plain. The characteristics of the Buckthorn family are—calyx of four or five sepals, and petals of the same number, and stamens answering to the sepals; fruit hard and dry. The common Buckthorn has serrated lance-shaped leaves, compact clusters of greenish flowers, and black bitter berries. The stamens are on one flower, and the pistils on another. The fruit gathered unripe yields a yellow dye; a coarse medicine for dogs is made from the ripe berries.

Edward was told that the Alder Buckthorn (R. frángula, *Plate VI.*, *fig.* 1), grew in a certain wood near Hawkhurst. He went to seek it, and was greatly disappointed to find that the whole wood, excepting the large trees, had been levelled to procure Hop-poles. However, he found it afterwards in the Bedgebury woods. It is a pretty shrub, with light green foliage; the broad-shouldered leaves are on long footstalks, and the little waxy pink-tinted flowers look very innocent in their small clusters. One of those caterpillars which children call "woolly bears" was crawling upon the leaves when Edward found it, and his young companion, a juvenile "man of Kent," told him by no means to touch the creature, for it was called a "black man's ring," and if it once got round his finger nothing could loosen it except cutting off the finger. Mr. Johns states that the caterpillar of the sulphur butterfly feeds principally upon this shrub, so probably it is a general favourite with caterpillars, and the "woolly bear" that did *not* fasten round Edward's finger, may disport among the branches next summer on gorgeous wings.

Of the same order is the Jujube Tree, a native of Syria, from the juice of which the cough lozenges called jujubes are made. At Hawkhurst Edward saw a graceful tree with pinnate leaves and long spines; it is an ally of the Buckthorn order, and is supposed to have been the tree from which our Saviour's crown of thorns was made. It is indigenous to Syria and Palestine.

We now come to a very important family, the Pea and Bean tribe. The British members of the group are characterised by two general features—a butterfly-shaped corolla (papilionaceous), and a *leguminous* seed-vessel. The *legume* is a pod, sometimes straight and sometimes curved; it prevails in the entire order. The papilionaceous corolla holds good in the British species, but it is wanting in some of the foreign ones. This order contains above six thousand species, most of them nutritious—all, according to Linnæus, wholesome. They are found in all parts of the world, with the exception of one or two remote islands, and are used for food for human beings as well as cattle in all countries. The timber trees Acacia, Rosewood, and Logwood, belong to this family, whilst gum arabic, senna, liquorice, and balsam of tolu are the produce of other species. Indigo and the Locust Tree also pertain to this group. The best gum arabic is the produce of an Arabian species of Acacia, and Arabian physicians first discovered the medicinal properties of senna. All the British species are herbaceous, with the exception of the Broom and Green-weeds. The leaves of this order are mostly pinnate, with a tendril or an odd leaflet, seldom simple. They have stipules in pairs. The flowers are handsome, often very sweet. The herbage is wholesome for cattle, the seeds for man. The "husks that the swine did eat" are generally considered to be those of the Locust Tree, sometimes called St. John's Bread, from an idea that its fruit was the locusts eaten by him; but as locusts have from time immemorial been the food of people in the East, the word may fairly be taken in a literal sense. The husks are still employed for feeding cattle in Palestine, and a juice is expressed from them and used in preserving fruits. The leaves and bark are used for tanning skins. They have ten stamens joined in two bundles or brotherhoods, and one pistil. This great leguminous order is divided into three groups—the Lotus group, the Vetch group, and the Jointed Vetch group. In the first, or Lotus group, *all* the stamens are united at the bottom.

The Furze is the first family of this group. Our common Furze needs no description, we have all seen it in every county we have visited, however far inland, or however close upon the sea. The French call the plant Jonc-marin, in allusion to its endurance of the sea-air. Its Latin name, Úlex européus (*Plate VI., fig.* 2), signifies *sharp;* and the fingers of eager flower-plucking children attest to the fitness of the appellation. This shrub is essentially characteristic of England; it cannot live in hot countries, nor does it extend far north, being scarce even in the Highlands of Scotland. When Linnæus first beheld the Furze, with its wealth of golden blossoms, he fell on his knees, and thanked God for its beauty. He attempted in vain to introduce it into Sweden. It perished in the open country, and even in the garden it sickened and died. Dillenius, too, another famous botanist, said that he could not find words to express the pleasure which the sight of this plant had given him. In Scotland it is called Whin more than Furze, and we often hear it named Gorse. Fences are made of this bush, the young shoots are eaten by cattle, the seeds are devoured by birds, and the old stems afford winter firing for the poor. At one time the whole plant used to be crushed by a machine and given both to cows and horses for food, in the neighbourhood of Birmingham.

There is a smaller species (Úlex nánus), with paler blossoms, which blooms in August and September. It grows in Kent and Sussex. Edward has specimens from the hills at Hastings; indeed he gathered some on the site of the famous battle which gave to William the First his title of "Conqueror."

The Green-weed family come next. The largest of these, the Dyer's Green-weed (Genísta tinctória, *Plate VI., fig.* 4), is pretty common. I have gathered it near Sutton, in the Ripon district, and in the neighbourhood of Richmond, and I see that Edward has specimens from Kent. It is a hard-wooded plant, but grows half prostrate, mingling with the grass in mountain meadows. Its large yellow flowers, placed

G

alternately up the stalk, and relieved by the smooth, dark, fine green foliage, render the plant both gay and elegant. It flowers in July.

The Needle Green-weed (G. ánglica), I found at Brimham Rocks—a wonderful collection of masses of millstone scattered upon a wide swampy moor near Harrowgate. This plant is almost as prickly as the Furze, it grows nearly prostrate, and its flowers are paler and smaller.

The Hairy Green-weed (G. pilósa), is destitute of prickles, its blooms are of a fuller shade, and are not clustered as in the Needle Green-weed. My specimen was sent to me from Hampshire. Once I saw it grow abundantly, but it was on sandstone rocks about Heidelberg, and those specimens are not admissible in our British collection.

The Dyer's Green-weed is sometimes used for imparting a yellow colour to wool.

The BROOM (Cýtisus scopária, *fig.* 3), is our well-known representative of the next family. More widely distributed than the Furze, but not growing in such immense masses, it is at once a handsome and a graceful shrub. Burns testifies his preference for it.

> "Their groves of sweet Myrtle let foreign lands reckon,
> Where bright beaming summers exalt the perfume;
> Far dearer to me yon lone glen o' green brecken,
> Wi' the burn stealing under the lang yellow Broom."

This plant used to be included in the last family. The name, Genista, was taken from the French—" Plant à genêt," or "genêt à balai," for from it they make brooms. Fulke, Earl of Anjou, being ordered to make a pilgrimage as a penance for his sins, and, at the same time to submit to castigation, wore afterwards a sprig of the rod in his helmet as a token of his humility, and hence received the name Plant à genêt. Geoffry, another Earl of Anjou, the husband of the Empress Matilda of Germany, went into battle with the same emblem on his helmet, and thus the name came to be bequeathed to his descendants the Plantagenets. In 1234, Louis of

France made the Broom the insignia of a new order of knighthood, and the members of the order wore a chain of the flowers, from which in allusion to Earl Fulke's reason for wearing it, a gold cross was suspended with the inscription "Deus Exaltat Humilis."

The Garden Laburnum belongs to this family; its seeds are an exception to the general rule of the tribe, for they are violently emetic, if not actually poisonous.

The common Rest Harrow (Onónis spinósa, *Plate VI., fig.* 5), is a shrubby, hairy plant, with simple leaves, serrated towards the end. The branches bear thorns, and the whole plant is so tough as to be said to arrest the harrow. Some go so far as to assert that this is the "thorn" spoken of in the curse of our first parents, but I see little reason for adopting the opinion. Its flowers are a brilliant pink; it flourishes on heaths and commons from June to August. My specimen was gathered near Ripon.

The small annual Rest Harrow is a very scarce plant, growing only on cliffs in the Mull of Galloway; it has pale lilac flowers, and leaves composed of three leaflets.

The useful Medicks come next in order. The Black Medick (Medicágo lupulína), is our first specimen. It grows abundantly among low herbage, and about the entrances of corn fields; as children we used to call it "Brimstone." It has small yellow flowers, packed closely into an oval cluster, and its leaflets are shaped like eggs upside down. Its legumes are black.

Our second species is the Spotted Medick (M. maculáta), which Fanny brought from Cornwall. It has only two or three flowers in a cluster, and its leaflets have a dark spot on them: hence the name. The seed-vessels are curious and beautiful, twisted round so as to form a ball, and the edges sharply toothed, so as to resemble a bur. On this account the plant is very mischievous in sheep pastures, for the legume gets entangled in the wool, and cannot be taken away without breaking the hairs. A Cornish merchant told me that the plant prevails to a great extent in Australia, and so damages the fleeces of the sheep that the wool from such estates sells for a much lower price than that which is free from burs.

Edward has brought the Purple Medick, or Lucerne (M. satíva, *Plate VI.*, *fig.* 6), from Kent. It is a much prettier plant than either of its brethren, its lilac flowers being arranged in a large cluster upon its tall leafy stem. It has also spiral seed-vessels, and is much cultivated for cattle.

The Sickle Medick is scarcely distinct from this, but its flowers vary to yellow and greenish-white.

There are two small members of this family, the Little Bur Medick, and the Toothed Medick, but we have not found them.

We now come to the Clovers or Trefoil family, to which the Melilot belongs; but, as some separate the Melilots into a distinct family, we will discuss them first. The name Melilotus signifies Honey Lotus. This division of the Trefoil family have their flowers in spikes, and their short legumes contain only one seed; the leaf comprises three leaflets. The common Melilot (Melilótus officinális), used to grow abundantly in the fields which now form the park belonging to the Palace of the Bishop of Ripon. I gathered the present specimen on some waste ground near Starbeck Station, and I have often seen the plant growing on the side of railway cuttings.

The White Melilot is much scarcer than the Yellow, but frequents the same kind of places. I have never found it.

The White Clover (Trifólium répens, *fig.* 7), is a very interesting plant, not only because of the valuable food which its blotched leaves afford to cattle, but because it is the ensign of Ireland, as Moore sings—

> "The Shamrock,
> The green, immortal Shamrock;
> Chosen leaf
> Of bard and chief,
> Old Erin's native Shamrock."

But far more touching than any allusion of Moore is the simple story of how the Shamrock became identified with Ireland. It was in the distant period when St. Patrick was preaching the Christian religion, and instructing the ignorant but intelligent people in the nature of the true God. When he asserted the doctrine of the Trinity, they would hear him no longer; hitherto they had listened with eager curiosity, but this mystery was too great for them to endure. "Three Gods and yet but one God, impossible!" they said. He stooped, and gathering a Shamrock leaf held it aloft. "Here is but one leaf," he said, "and yet, lo! it is three! If in one of the humblest works of creation this mystery is manifested, how much more should it exist in the Creator himself!"

> "Full well he read, that mighty man of old,
> A mighty mystery of the humble sod;
> With wondering awe they saw the saint unfold
> The triple leaf, and teach a triune God!
> Then unbelief and prejudice took flight,
> With such 'weak things' did God the wise confound;
> And darkness fled before the flood of light,
> And heathen ears received the Gospel sound."

The next neighbour of the White Clover, the Bird's-foot Clover (T. ornithopodoídes), has a small pale red flower, about three in a cluster. It is said to grow at Musselburgh, but I have never seen it.

The Pink Clover (T. praténse), which, as children we used to call "Sweet Kitty Clover," is common in meadows everywhere.

The Zigzag Clover (T. médium), is a taller plant, with longer leaflets, and a less dense head of flowers. I gathered it near Goostrey, in Cheshire, by the side of a sandy lane, and I see Edward has a specimen from Kent.

The Sulphur Clover (T. ochroléucum), with its erect downy stems and dense head, favours dry pastures in the eastern counties, but I have not had the good fortune to procure it.

The Hare's-foot Trefoil (T. arvénse), I found in a lane leading west by the side of the Ure, near Ripon. The calyx-leaves are long and densely hairy; and the small flowers are almost lost among them. It blooms also in the autumn among corn.

The Teasel-headed Trefoil (T. marítimum), has a striking resemblance in miniature to the plant after which it is named, It inhabits salt marshes, and we have no specimen.

The Hop Trefoil flowers in compact yellow heads; it spreads widely, and is a gay little plant. I gathered it in a Clover field in Swaledale.

The Lesser Trefoil and the Slender Trefoil (Trifólium mínus and filifórme), are of the same habit as the Hop Trefoil (T. procúmbens), but very minute. I gathered these plants on the sandy pastures at Redcar, on the east coast of Yorkshire.

The Subterraneous Trefoil (T. subterráneum), is one of Fanny's Cornish treasures; it grows half buried in sandy soil, but the clusters of four flowers manage to raise their heads a little, and look pretty on account of the strong contrast between the crimson throat and white wings of the corolla. The calyx throws out fibres when the flower has faded, by means of which it drags the plant down and buries it.

The Rough Trefoil (T. scábrum), is from Polperro, on the same coast. The calyx-leaves are naked and spreading, and give the clusters a starry appearance; these grow both at the ends of the branches and from the axils of the leaves. The flowers are pale pink and very small.

The Strawberry-headed Trefoil (T. fragíferum), grows on the salt marshes between Clevedon and Weston-super-Mare; its round cluster and small pink flowers appearing from among the teeth of the calices, give it a strong resemblance to a half-ripe Strawberry. Fanny sought the Knotted Trefoil on the spot indicated by Mr. Johns, in his "Week at the Lizard," but in vain.

The characteristics of the Bird's-foot Trefoils are a five-cleft tubular calyx, and wings the same length as the standard. All the members of the family are yellow. Edward has a beautiful specimen of the Greater Bird's-foot Trefoil (Lótus májor), from the Kentish lanes, and I have them also from Yorkshire ones. The cluster of yellow flowers is large and handsome, and the plant grows to the height of two or three feet; it is covered with soft hairs.

The Bird's-foot Trefoil (L. corniculátus, *Plate VI., fig.* 8), is a cheerful inhabitant of every lane and field in nearly every county. Its pretty sweet-scented yellow clusters, tinged with crimson when in bud, are welcome everywhere; its strong deep-penetrating root makes a secure anchorage even in the shifting sand of the river-bed, and the equally unreliable bank; it covers plots on the chalk downs with its bright flowers, and adorns the exhausted lime-quarry; indeed the difficulty would be to say where it does *not* grow.

The Slender Bird's-foot Trefoil (L. tenuíssimus), Fanny found at Clevedon; it has only two or or three flowers in a cluster.

The Smallest Bird's-foot Trefoil (L. angustíssimus), grows on cliffs about Talland and Polperro; it has only one flower on a stalk, or occasionally two. The plant is very woolly.

We have none of us found the Oxytrópis; but I have a pretty specimen which was sent to me from Scotland. The large blue flowers are collected in heads, and open in July (Oxytrópis uralénsis).

The Lady's-Finger (Anthýllis vulneráría), is a plant with woolly foliage. Its flowers are yellow, and grow in pairs of dense clusters. A general observer would call it a well-grown yellow Clover, but upon a closer inspection you at once perceive that the large head is composed of two distinct clusters. This plant grows commonly in pastures about Ripon. I have also found it in the vicinity of Richmond, and there are specimens here from Kent and Warwickshire.

The Milk Vetches have tumid legumes, white or purple flowers, and pinnate leaves.

The Sweet Milk Vetch (Astrágalus glycyphýllus), grows two or three feet high when supported, but more often trails along the ground, or among low bushes. It has large, light green leaves, composed of nine or more oval leaflets; its flowers are greenish-white, and grow in a large cluster; its stalks are sweet, with the flavour of liquorice, and children like to nibble them. It grows abundantly about Frome, in Somersetshire, but I made my first acquaintance with it in the woods of Switzerland.

The sandy pastures about Redcar produce the Purple Milk Vetch (A. hypoglóttis, *Plate VI., fig.* 9). It has a large cluster of purple flowers, nearly as big as the rest of the plant. The leaves are pinnate, the leaflets small, but very numerous.

We come next to the true Vetches. The most beautiful of them, the Wood Vetch (Vícia sylvática, *Plate VI., fig.* 10), grows in the hilly parts of Yorkshire and Herefordshire. By means of its delicate tendrils it climbs the bushes as high as six or seven feet, and from thence its light graceful foliage, and large clusters of pale, purple-striped flowers, hang in elegant festoons. It blooms late in the summer.

The Tufted Vetch (V. crácca), is the next in order of beauty. It climbs the hedges, or roams over the bank, raising its one-sided spikes of bright blue flowers wherever it can find a support. I have gathered it in many parts of Yorkshire, Lancashire, Cheshire, Warwickshire, and Kent.

The common Vetch (V. satíva), is cultivated as fodder for cattle. It is a succulent herb, with a thick stalk; its flowers, growing alone or in pairs, are seated on the main stem.

The Narrow-leaved Vetch (V. angustifólia), resembles the common one, but is of a more slender habit. I found it first on the ruins of Jervaulx Abbey, in Wensleydale; and since, I have seen it growing in Edward's favourite county, Kent.

The Bush Vetch (V. sépium), with its grey purple clusters at the axils of the leaves, is a very familiar ornament of our heaths and woods, flowering in spring.

There are a Sulphur Vetch and a Yellow Vetch, both frequenting the south coast of England, and a Purple Vetch, with only two pairs of leaflets on each leafstalk; but none of these have rewarded our researches.

The Tares are pretty slender plants. The Hairy one (Érvum hirsútum), very elegant, with its pale lilac-striped blossoms on long slender footstalks, alone or in pairs. This plant grows abundantly in the South. Fanny has Cornish specimens, and Ned Kentish ones, but I have not found it in the North.

The Smooth Tare has very insignificant blossoms, three or four in a cluster, but its foliage is very light and elegant, and has a pretty effect climbing among the tall plants in a hedgebank. It abounds in every part of England.

The common Bird's-foot (Ornithópus perpusíllus, *Plate VI.*, *fig.* 13), has hardly larger blossoms, but they are tinged with bright rose colour, which makes them more attractive. The plant is compact, the numerous leaflets in the pinnate leaves are oval and pointed, and the seed-vessels are curved, and knotted like a string of beads. I have never found the plant but in the sandy lanes of Cheshire.

The Horse-shoe Vetch (Hippocrépis comósa), allied to the Bird's-foot by its curved beaded legumes, is a trophy from the downs of Wiltshire. Its clusters of yellow flowers nearly resemble those of the Bird's-foot Lotus, but the more numerous leaflets and peculiar seed-pod mark a broad distinction between the plants. Both flower in June.

On one of the downs bordering upon Salisbury Plain I got the beautiful Saintfoin or French Hay (Hedýsarum onóbrychis). It is largely cultivated in some districts for cattle, but as Salisbury Plain is given as one of its natural habitats, I hope mine was a wild specimen. It is a beautiful plant, with a dense spike of crimson and white flowers, and bright green pinnate leaves. Ann Pratt speaks of a foreign species of Saintfoin which keeps up an incessant motion (H. gýrans). If the leaflets are held quiet by force, they move afresh upon being released with an increased rapidity, as if to make up for lost time.

The Everlasting Pea family is the next one in this large

group which we shall discuss. There are fewer leaflets in the members of this family, and each leaf is endowed with a tendril. The handsomest species is the Broad-leaved Vetchling (Láthyrus latifólius). This is a frequent ornament of gardens and shrubberies, only objectionable because of spreading too quickly. Its large crimson standard, as the uppermost petal in the papilionaceous order is called, and purple-tinged wings, are very conspicuous. It is very rarely found wild.

The Sweet Pea is a member of this family, and is indigenous in the south of Europe, especially Sicily. Keats greatly appreciated its beauty, likening its flowers to the gay insects which they resemble in form.

> "Here are sweet Peas on tiptoe for a flight,
> With wings of gentle flush o'er delicate white;
> And taper fingers catching at all things,
> To bind them all about with tiny rings."

The Field and Marrowfat Peas are its near relations. The Field Pea produces the "dried Peas" of the shops, from which we make "pease pudding." The Marrowfat Peas are boiled green for the table, and the young pods make excellent soup.

Fanny describes the Narrow-leaved Vetchling (Láthyrus sylvéstris), as drooping over cliffs at Looe, and, in particular, adorning a rocky path by which they descended to some caves, among a group of rocks called the "Quakers' Meeting," which was the "bathing place." After the refreshing bath they were accustomed to twine the Vetchling round their hats as they sat in the sunshine eating the biscuits rendered necessary by the appetising air and ablution. In this plant the flowers are large, the standard greenish, and the wings a purplish-crimson.

The Blue Marsh Vetchling is a rare plant, as is also the Rough-podded Vetchling, with its bright crimson standard. We have no specimen of either.

I found the Yellow Vetchling (L. áphaca), upon the railway line near Warminster; it has glaucous foliage, large heart-shaped stipules, and tendrils without leaves. The flower is a pale yellow. I have since found it on field borders in the same neighbourhood. It is a smaller plant than the Narrow-leaved species. Its seeds produce headache, so it must be placed with the Laburnum, as an exception to the general rule of the wholesomeness of leguminous plants.

The Crimson Vetchling (L. nissólia, *Plate VI.*, *fig.* 11), is a very elegant plant, growing about a foot high, with a slender stem, awl-shaped stipules, and grassy leaves without tendrils. The flowers are solitary, upon a long footstalk, and both standard and wings are of a bright crimson. Edward found it in a lane near Hawkhurst last June, and it also grows near Frome, in Somerset.

The Meadow Vetchling (L. praténsis), is a familiar hedge ornament, decking with its full clusters of bright yellow blossoms groves and hedgerows in every county I have ever visited.

The Sea Pea (Pisum marítimum), has large flowers in compact clusters, both standard and wings are purplish-crimson; each leaf boasts eight pairs of glaucous leaflets, and its stem is prostrate. It frequents stony shores.

The Bitter Vetch is the last leguminous family. These plants have no tendrils, and do not grow more than a span high.

The common Bitter Vetch (Órobus tuberósus, *Plate VI.*, *fig.* 13), has crimson blossoms turning purple as they fade. It is common in woods everywhere. In the Highlands of Scotland it is very abundant, and the people eat the root, or steep it in their liquor. It has somewhat the flavour of liquorice, and they believe that it prevents hunger. They call it "Corneille" or "Heath-pea." It is an early flower, and both its leaf and blossom are desirable in the spring nosegay.

The Black Bitter Vetch and the Wood Bitter Vetch (O. níger and sylvática), are rare Scotch species; the former has blue flowers, the latter cream-coloured ones.

We have now gone through this very important order of the

leguminous plants. We have observed their wonderful structure, the stamens and pistil, the fruit-bearing parts protected by two petals united and shaped like a *keel;* two more petals, placed one on each side of the keel, called *wings*, which further protect the delicate parts; and above all a *standard* petal, broad and sheltering, which, by a little twist of the footstalk always turns to the wind, and beneath the bitter blast bends and covers the rest of the flower, preserving the germ of future life.

Thus, this tribe alone would reveal to us the tender care and great wisdom of our Creator and Father. The mess of pottage, for which the impatient Esau sold his birthright, was made of a member of this tribe—the Lentil (Ervum lens); and the food which nourished the four faithful Jews in the court of Nebuchadnezzar, and made their countenances fair and flourishing, was Pulse, another plant of this family. Thus the leguminous order convey both warning and example; warning against an inordinate and impatient desire, and example of the blessings of self-denial and moderation, when practised from a holy motive and consecrated by a pure aim.

CHAPTER VII.

ROSÁCEÆ.

> "Where do the Wisdom and the Power Divine
> In a more bright and sweet reflection shine?
> Where do we finer strokes and colours see
> Of the Creator's real poetry,
> Than when we with attention look
> Upon the third day's volume of the book?
> If we could open, and extend our eye,
> We all, like Moses, should espy
> E'en in a bush the radiant Deity."
>
> COWLEY.

THE ROSE order, like the Leguminous, requires to be divided into groups, distinguished from one another by the form and structure of the fruit. A four or five-lobed permanent calyx, five petals, and numerous stamens, are the characteristics of the whole order. A simple seed or kernel enclosed in a hard case, and surrounded by juicy pulp, distinguishes the *Almond* group; a collection of seed-vessels opening at the side distinguishes the *Meadow Sweet* group; numerous carpels, fastened into a receptacle either fleshy or dry, characterise the *Strawberry* group; seeds enclosed within a hardened calyx mark the *Burnet* group; nut-like seeds in a fleshy tube the *Rose* group; while, in the *Apple* group, the seeds are contained in horny cells.

To begin with the Almond group. The first family is that of the Plum and Cherry.

The Wild Cherry (Prúnus cérasus, *Plate VII., fig.* 10), abounds in Kent; it is a beautiful object when covered with its sheet of snowy blossoms, and scarcely less so in its autumnal garb of deep crimson foliage. It grows very abundantly in the woods about Richmond also. There are two varieties—one bearing red and one black berries; but between the birds and the boys no fruit has been left to ripen upon the trees within my range of observation. But never have I seen such Cherry-blossom as in Switzerland, and that blossom was followed by abundant and well-ripened fruit. There they distil a strong coarse spirit from the Cherries, called Kirschwasser, or Cherry-water. We were often disappointed when, stopping at a wayside inn on a long excursion, we asked for wine or milk, and could get nothing but this abominable spirit. But we had our enjoyment out of the Cherries; for one penny we could buy a basketful, and we regaled ourselves with them daily. At Hamburgh an annual feast of Cherries is held, the occasion of which is interesting. When the city was besieged by the Hussites, and reduced to great straits, the suffering people dressed their children in mourning and sent them to sue for mercy. The enemy, touched by this sight, fed them bountifully and sent them back with branches of Cherries. Processions of children celebrate the annual feast. Mr. Johns gives an interesting account of this in his " Forest Trees of Britain."

In the same Yorkshire woods the Bird Cherry (P. pádus), puts forth its spikes of elegant flowers, and this tree is still more abundant in the glens in Swaledale. The fruit is dark purple, oval, and bitter to the taste.

The Common and Portugal Laurels belong to this group, also the Peach, Nectarine, Apricot, and Almond. The fruit of all is wholesome, but the leaves contain prussic acid in a greater or less degree; and this is, as you know, a deadly poison. The bark yields gum. I remember a beautiful Morello Cherry tree, trained against a south wall in our fruit garden, parted with so much gum that the soil was caked with it. The tree died of exhaustion.

Kent has supplied us with specimens of all the wild Plums.

The common Wild Plum (Prúnus doméstica), is distinguished by being thornless.

The Wild Bullace (P. insitítia), has thorns at the end of the branches.

The Blackthorn or Sloe has abundance of thorns, hence its Latin name (P. spinósa). This species is the earliest in flowering, its abundant blossoms appearing before the leaf-buds, in March, about the time when the east winds set in. On this account a cold spring is called a "Blackthorn winter." In Wiltshire I remember finding bushes in the hedgerows covered with the keenly-acid blue-black fruit of this shrub. They were beyond my reach, so I called a labourer working in the field, and asked him to gather some for me. It was long before I could make him understand which object in the hedge had excited my desire; and when at last he comprehended me, he exclaimed in supreme contempt—"I could not go for to think that you could be a wanting of snags!" Such is the provincial term for the Sloe. The French adulterate port wine with this fruit, and also pickle it green, like Olives.

The Meadow Sweet group (Spiræáceæ), contains but one British family. The common Meadow Sweet (Spiræa ulmária), is easy to find, its powerful scent being perceived from afar, even were its clusters of numerous white feathery flowers less conspicuous. The French call it "Reine des prés," and its excessive fragrance makes it indeed worthy to be a queen. But the powerful scent is not wholesome, for I have read of children who have kept a quantity in their little bedroom being found insensible in the morning. The other members of this family call themselves Dropworts. They are both of them scarce.

The Common Dropwort (S. filipéndula), grows on the serpentine cliffs in the Lizard district; Fanny got specimens from thence last July. It is a smaller plant than the Meadow Sweet, and has little scent; its leaves are more or less pinnate—not merely deeply serrated as in the former plant.

The third Spiræa attains the stature of a shrub; its leaves are simply lance-shaped, evenly serrated, and smooth. Its flowers are pink, and grow on a smaller, more regular cluster than those of its brethren. These specimens come from the neighbourhood of the English Lakes.

The Willow-leaved Dropwort (S. salicifólia), is rare in its wild state, but frequent in shrubberies.

This specimen of the White Drýas (D. octopétala), the only representative of the first family in the Strawberry group (Dryadeæ), was sent to me from the Perthshire Highlands. Its large blossom has eight petals of brilliant whiteness, and its simple oak-shaped leaves are white and downy beneath. It flowers in August. It is named from the Greek word for *oak*, on account of the shape of its leaves.

The two members of the Avens family are also in my possession.

The common Avens or Herb Bennet (Géum urbánum), opens its yellow flowers, disclosing the cluster of crimson stamens and stigmas. The petals fall off as soon as gathered, and the only chance of pressing a piece satisfactorily is to make it bloom in water, and transfer it at once to the paper. The leaves at the root are pinnate, those of the leaves ternate, or in three leaflets. The seeds are fastened to the receptacle, and each is endowed with a feathery plume.

The Water Avens or Long-Bennet (G. rivále, *Plate VII.*, *fig.* 1), is a deep orange; the petals form a bell, and the full flower always droops; it grows by rivulets and in moist woods. I remember seeing it for the first time in the Mackershaw woods, near Ripon, a part of the estate of Studley—a lovely situation, such as might well suggest the aspiration of the poet Browne—

> "Oh blessedness to lie
> By the clear brook where the Long-Bennet dips."

The distinguishing characteristic of the Cinquefoil family, is, as the name indicates, the leaf of five leaflets. But to this characteristic there are exceptions.

The Silver Weed (Potentílla anserína), is our most familiar Cinquefoil, yet the shining leaves which adorn our roadsides are pinnate. It has large, light yellow flowers; and its creeping stems and shining white foliage make it a general favourite.

The Shrubby Cinquefoil (P. fruticósa), has a woody stem, and grows to the height of three feet. Its leaves are smaller than those of the Silver Weed, and are dark green on the upper side, but white and woolly below; they likewise are pinnate. This is a scarce species, its bright rose-shaped yellow flowers are thickly scattered over the bush, and it is a pleasing object. I gathered it early in last August about rocks in Teasdale, not far from the celebrated waterfall called "The High Force."

The Strawberry-flowered Cinquefoil, which has occasionally seven leaflets, I have not found. It only flourishes in Wales.

The Spring Cinquefoil I gathered at the end of last April on Blackford Hill, near Edinburgh. It is a very small species, with yellow flowers, and deep green leaves. The edges and the ribs of the leaves are somewhat bristly; otherwise the plant is smooth.

The Creeping Cinquefoil (P. répens, *Plate VII.*, *fig.* 2), grows in most hedgebanks; its trailing stem, leaves and flower-stalk joined to the stem in clusters, bright yellow blooms, and full green five-parted leaves, are easily recognised.

The Hoary Cinquefoil grows also on Blackford Hill, but I was too early for it. It is a cottony plant, with small yellow flowers.

The Hairy, the Alpine, and the White Cinquefoils are mountain plants. The Alpine species is orange; the Hairy, bright yellow.

There are two members of this family with ternate leaves—the Strawberry-leaved Cinquefoil (P. fragariástrum), familiar to us all, but generally mistaken for Wild Strawberry blossom; and the Three-toothed Cinquefoil, each leaflet of which has three points. This last is a Highland species.

The Strawberry-leaved Cinquefoil may be distinguished from its namesake by its earlier period of flowering, weaker habit, and, most conclusively, by its naked seeds, placed upon a dry receptacle.

This family have no uses that I am aware of; boys sometimes eat the roots of the Silver Weed, and say that they resemble Chestnuts when roasted. Many garden species of Potentilla have large richly-tinted blooms and silvery foliage, and form handsome border plants.

H

The Tormentil family succeeds that of the Cinquefoil. There are only two British species; these are distinguished by having four petals, while the Cinquefoils have five.

The common Tormentil (Tormentílla officinális), is a frequent ornament of our pastures and moors; its flowers resemble that of the Spring Cinquefoil, except in the number of its petals.

The Trailing Tormentil (T. réptans), is rare, but my sister found it in Cumberland last autumn. It is very like the common species but larger, its stems longer and more prostrate, and its flowers of a greater size.

The inhabitants of the Hebrides and Orkney Islands use the roots of the common Tormentil for dyeing, preferring them to oak bark. So much land has been destroyed by digging for these roots, that the practice is now prohibited in some of the islands.

The Sibbaldia has a family to itself. It is a thick, bushy plant, with prostrate stems, ternate leaves, and clusters of tiny yellow flowers. It grows in the Highlands, and does not yet grace our collection.

The Marsh Cinquefoil (Cómarum palústre, *fig.* 3), is larger and stouter than most of the true Cinquefoils. It has dark purple flowers, and its large sepals are tinged with purple. I found it when I was staying in the vicinity of Victoria Park, Manchester. We were walking to visit some dyeing mills, and passed by fields and waste ground. I espied a pond in one of these fields, and the fence being easy to climb I surmounted it, telling my friends that I would meet them at the further corner of the field. They watched me as they would have done a Zoolu, or a Red Indian; while I, unconscious how strange my conduct might appear to townspeople, wandered round the pond, gathered the Marsh Cinquefoil on its margin, and returned well satisfied to my companions. I was immediately overwhelmed with questions—"What do you want that ugly flower for? How did you know it grew there?" "What is the use of such things?" It was impossible to convey to their minds the intense pleasure which a new specimen afforded me, nor to justify my habit of examining every pond that came in my way,

We now come to the Strawberry family, and here are the only two British species.

The Wild Strawberry (Fragária vésca, *Plate VII.*, *fig.* 4), needs no description; its fruit and flower are familiar to us all from childhood. They are found in every country, and flourish on crumbling rocks and venerable ruins, as well as under sheltering hedgerows, or in sunny groves.

The Hautbois Strawberry (F. clátior), grows in the woods near Clevedon and in Herefordshire; Fanny found it at the beginning of June. She was attracted by its large size, and the spreading hairs on its stalks proved it to be the species in question. Some children who came to stay with her later in the season went to her in great glee, to show what large wild Strawberries they had found in the woods. She says that one flower in the cluster she gathered had no pistils. This variability in the bloom is a distinctive mark of the species.

The Brambles, which are at once the grief and joy of children, are the next family in the Strawberry group. The rose-shaped flower has crumpled petals, and thick powdery-looking sepals. The fruit is a collection of seeds covered with juicy flesh; these are called *drupes*. The leaves are formed of either three or five leaflets. The Latin name is from Ruber, red.

The common Blackberry (Rúbus fruticósus), called by Yorkshire country people "Bumble Kite," is abundant everywhere.

> " Thy fruit full well the school-boy knows,
>   Wild Bramble of the brake;
> So put thou forth thy small white rose,
>   I love it for thy sake.
> How delicate thy gauzy frill,
>   How rich thy branching stem!
> How soft thy voice when woods are still,
>   And thou sing'st songs to them.

> "The Primrose to the grave is gone,
>   The Hawthorn flower is dead;
> The Violet by the mossy stone
>   Hath lain her weary head,
> But thou, Wild Bramble, back dost bring,
>   In all their beauteous power,
> The fresh green days of life's fair spring,
>   And boyhood's blooming hour."

There are several Brambles, which, like this, have angular stems. We have only one of these—the Red-fruited Bramble (Rúbus suberéctus). Here the fruit is deep red instead of black; it grows on the banks of a little stream in Swaledale. Several other so-called species differ almost imperceptibly from these, and, according to the opinion of many eminent botanists, should never have been called by different names.

The Wild Raspberry (R. idæus), grows luxuriantly at Grantley Lakes, near Ripon. These are pretty pieces of water surrounded by rocky woods; they lie off the main road, about two miles from Grantley Hall, the seat of the nobleman of that name. These lakes used to be a favourite place for pic-nics with us, and the beautiful wild Raspberries had their share of attraction. I saw them just as abundantly on the banks of Loch Katrine, and about Oban, a year or two ago. They are quite as sweet and as well-flavoured as the garden Raspberries, only decidedly smaller. The flowers of this species are pendulous. The Raspberry grows freely in Swaledale; and in the same localities, creeping by the side of the rocks, flourishes the Stone Bramble (R. saxátilis), an herbaceous plant, about a foot high, bearing a cluster of few flowers, and light red berries.

The Cloudberry (R. chamæmórus), I sought in the same district long in vain; but at last I found a farmer who spoke of curious orange berries gathered on low plants on the hills. I went to his farm, and climbed the hills behind it, and in due time I found, among the Ling on the summit, the simple plaited leaves and single flowers of the herbaceous Cloudberry. The plant is without thorns, the flowers have seldom the stamens and stigmas in one bloom, and the grains of the fruit are large and tawny. It only grows where the clouds rest.

The Dewberry (R. cǽsius, *Plate VII.*, *fig.* 5), has a fruit of few grains of a bluish-black colour; these are covered with a bloom like that on a Plum. Its flowers are pure white or tinged with lilac.

The Hazel-leaved Bramble (R. corylifólius), has larger flowers in a more scattered cluster, and long rambling *round* stems; in this last particular it agrees with the Dewberry and Raspberry; the more near allies of the Bramble have *square* stems. Edward brought specimens of these from Kent.

We must not set aside the whole troop of Blackberries as mere play-feasting for children; an excellent jam is made of the fruit, which is pleasant, wholesome, and even medicinal. Notable housewives make wine from the berries both in France and England; in the former country the bush is called "Pinte de vin," in honour of this use. The young tops and stems were eaten as salad by the Greeks, and are still used in dyeing.

The Bramble was the subject of one of the earliest parables of Palestine; and by analogy, with it the wise young man Jotham showed to the men of Shechem the folly of anointing a weak king to reign over them. The Bramble, being a trailing plant, cannot give support to another, but needs a prop for itself. It would thus suggest the far higher lesson, not to "trust in princes, or in any child of man, in whom is no strength;" but to "trust in the Lord for ever, for in the Lord JEHOVAH is everlasting strength."

The Agrimony family is now the only one that remains of the Strawberry group. Our one British species (A. eupatoria), grows in lanes near Ripon, and near Richmond; it has five petals, five sepals, and two seeds enclosed in the tube of the hardened calyx. Its yellow flowers grow on a tall slender spike, and have an aromatic smell. It blooms in July. It used to be valued as a tonic medicine.

The Burnet group comes next. The seeds are contained in a hardened calyx, but the flowers have the peculiarity of being endowed with two calices, and are entirely destitute of corolla.

The little green clustered blossoms of the common Lady's Mantle (Alchemílla vulgáris), are familiar to us all; it grows in meadows and pastures, attains the height of one foot, and bears its flower-cluster on the summit of its branches.

The Alpine Lady's Mantle (A. alpína), was sent me from the Sma' Glen, Perthshire. Its flowers are no more conspicuous than those of the former species, but indeed less so; its great beauty consists in the glossy green of its palmate leaves, contrasted with their silver lining.

The Field Lady's Mantle (A. arvénsis), is in Edward's collection of Kentish plants; it grows abundantly in corn fields in that and most other counties—an humble, unassuming plant, with flowers and leaves of the same dull light green hue. It has large, broad, jagged stipules.

The tall Burnet (Sanguisórba officinális), grows in meadows by the Swale, near Richmond. It has a pinnate leaf, and a very tall flower-stalk, surmounted by a compact oval head of dark crimson flowers. The leaves are furnished with stipules. This is our only British species.

The Salad Burnet family (Potérium sanguisórba), is the last in this Burnet group; it also has but one British member. This Salad Burnet grows in our Yorkshire pastures, the stamens hanging in long clusters from the tiny flowers of the head; the stigmas live on a separate plant. Its leaves have the flavour of Cucumber, and are, therefore, used in salads; they are also an ingredient in the beverage called "Cool Tankard."

Now we come to the Roses, the prettiest family in all the extensive order to which it gives the name. The five petals and permanent sepals are characteristics common to the whole tribe; but the enlarged and fleshy tube of the calyx, which becomes the seed-vessel, is a distinctive feature of this family.

The Burnet Rose (Rósa spinosíssima), is a dwarf species; the leaves are small, and every part of it bristly; the flowers are cream-coloured, and very fragrant. I have seen it growing abundantly on the shore of the Firth of Forth, and Fanny describes it as flourishing by the Loe Pool. It blossoms in July, and its dark brown berries ripen in September.

The Sweet Briar (R. rubiginósa), grows side by side with the Burnet Rose on the shore of the Firth of Forth, and it is also an inhabitant of Cornwall, Fanny having gathered it near Fowey. The fragrant scent of its foliage is better recognised as a mark of distinction than its frail pink blossoms. This is the Eglantine of the poets.

The Dog Rose (R. canína, *Plate VII., fig.* 6), is a very common species, adorning every hedge in July with its fragrant flowers. It has pink blossoms, and not bright foliage; and its oval scarlet fruit is well known. In Queen Elizabeth's time ladies used to make a conserve of these hips, and they are still used medicinally. From the flowers excellent Rose-water is distilled. The Romans strewed Roses in the streets at their great festivals, and the Egyptians made the Rose an emblem of silence: hence I imagine, the saying "Under the Rose." With us it is rather an emblem of love.

The Soft Rose, and the Downy-leaved Rose, grow in the woods in Swaledale; they resemble each other closely, the foliage of both being rough, and having a strongly resinous scent. The former species has large red fruit, tipped with the old sepals; the flowers and fruit of the latter are both darker in shade, and the form of the fruit is longer.

The Round-headed Rose, with ribbed fruit, and the small Sweet Briar, with pale blossoms, I have never found. The latter is said to grow at Bridport, in Warwickshire. The Glaucous Rose is a Highland species.

The Trailing Rose is a pretty delicately-scented species, common enough in our hedges, and easily distinguished by its bright pale foliage, and the salmon tint of its petals.

The Thicket and Irish Roses have hooked prickles. I have got no specimens of them.

The Trailing Dog Rose (R. arvénsis), is my last contribution. It is very common in Yorkshire, climbing a great height.

Its flowers are large, pure white, and very sweet. Its foliage is scentless, of a full bright green, and the large leaflets taper very gracefully. This is reputed to be the White Rose which became the standard of the Yorkists. I fancy it is also the species sometimes called Musk Rose, its scent partaking of that odour. It was a great favourite with Keats, who writes of its charms again and again—

> "I saw the sweetest flower that Nature yields,
> A fresh-blown Musk Rose; 'twas the first that threw
> Its sweets upon the summer: graceful it grew
> As is the wand that Queen Titania wields:
> And as I feasted on its fragrancy,
> I thought the garden Rose it far excelled."

In Scripture the Rose is spoken of as a type of prosperity: "The desert shall rejoice and blossom as the Rose." Dr. Murray states that the Rose used to be considered to possess medicinal properties, and enormous quantities of the Damascus Rose were cultivated in the East for chemical purposes: he speaks of a breakfast given at Hazar Bagh, or "thousand gardens," near Shiraz, to Sir John Malcolm, then our envoy to the Court of Persia, which was celebrated on a stack of Roses. "The stack, which was as large as a common one of hay in England, had been formed without much trouble from the heaps of Rose-leaves collected before they were sent into the city to be distilled." He relates, also, that a courier who brought despatches overland from Constantinople to the British Government was robbed and plundered of everything by thieves. Among the goods he was conveying were several bottles of attar of Roses, and, one of these breaking, the scent was so powerful as to guide the officers of justice to the robbers' den. Herrick draws a simple moral from the fast-perishing Wild Rose—

> "Gather the Roses while ye may,
> Old Time is still a-flying;
> And this same flower which smiles to day
> To-morrow will be dying."

and the same idea of the fleeting nature of all joys, of which the Rose might stand as emblem, appears in the pathetic lines of Mrs. Hemans:—

> "It is written on the Rose
> In its bright array—
> 'Hear thou what these buds disclose,
> Passing away!'"

We now come to the Apple group, in which the seeds are ensconced in a horny case or *core*, and further protected by the juicy swollen flesh of the calyx-tube.

For some years I sought the Wild Pear in vain, until some botanical friends of mine migrated to Shropshire, and then I got information of the object of my search. This tree is indigenous in that county as well as in the neighbouring one of Herefordshire—the great locality for Pear cultivation. Here I had afterwards the good fortune to find it growing in Penyard Wood, near Ross, and blossoming in April. A Pear, very little removed from the wild species, is there grown in the open fields, and called Barland Pear. The fruit is safe enough from depredation, for it is so woody, even when ripe, that it has originated a country proverb, "as hard as a Barland Pear." There Pear orchards exist as systematically as Apple orchards, and the perry—a refreshing beverage somewhat resembling cider—is made from the fruit. Before Kensington became so extensive, the inhabitants used to gather the Wild Pear blossoms in the hedges; but squares and palaces stand now on the home of the wild shrubs.

The wood of the Pear tree (Pyrus commúnis), is very hard, and is used instead of Box by some for wood-engraving.

The Crab Apple (P. málus, *Plate VII.*, *fig. 7*), is familiar to us all for its clusters of pale cupped flowers and rose-tinted buds. I have gathered it in a lane at the back of the Bishop's Palace, at Ripon; also in the Richmond district; and, abundantly near Horningsham, in Wiltshire. Fanny has specimens from Somersetshire, and Ned from Kent. The orchards of Herefordshire, Somersetshire, and Devonshire are beautiful objects when covered with blossom, and the busy scenes of the cider-making are not less interesting. The Apple is the most useful of British fruits, and a merry business the gathering generally is. Solomon had a great preference for this tree, as he says, " As the Apple among the trees of the wood, so is my Beloved among the sons;" and again, "A word fitly spoken is like Apples of gold in pictures of silver."

The Wild Service tree (P. torminális), is the next member of the Apple group. It grows wild in the woods about Clevedon; it has large leaves, with pointed lobes, and clusters of white flowers. The fruit is a small berry, brown, and spotted.

Here also flourishes the White Beam tree (Pýrus ária); its simple leaves, lined with white cottony down, are very conspicuous. It has large showy bunches of white blossom, and oval berries, red, and shading to brown. There is a group of these trees of a great height in Longleat Park; and, placed as they are in front of a Fir-grove, the effect of the white foliage is very striking. They grow also near a small lake on Lord Bath's property, called Shearwater, and in other parts of the same district.

The True Service tree (P. doméstica), I have a specimen of, but not a wild one. It greatly resembles the Wild Service, but has pinnate leaves, and whitish stalks.

The Mountain Ash (P. aucupária), is the glory of our hilly woods, hanging its long branches and gay clusters of red berries over rocks and cascades, and adding everywhere to the beauty of the landscape. Its pinnate leaves and white clusters of bloom are familiar as an ornament to the shrubbery. Bird-catchers in France and Germany bait their traps with these berries. This tree was highly esteemed by the Druids, and it is always found growing about the places they frequented. An old superstition existed that the Mountain Ash preserved from evil spirits, and on this account it was planted near houses, both in England and Scotland. In the Highlands it is still deemed lucky to drive cattle with a branch of

"Rowan," as they call the tree. The Welsh children term it "Wiggin;" they gather the berries, and the mothers make sweet liquor of them. The leaves of this tree contain prussic acid. Wordsworth speaks glowingly of the Mountain Ash:—

> "No eye can overlook, when 'mid a grove
> Of yet unfaded trees she lifts her head,
> Decked with autumnal berries that outshine
> Spring's richest blossoms; and ye may have marked
> By a brook-side or solitary tarn,
> How she her station doth adorn; the pool
> Glows at her feet, and all the gloomy rocks
> Are brightened round her."

The Medlar (Méspilus germánica, *fig.* 8), is the next family in the Apple group. I have a garden specimen of its one representative, but it is the wild kind. The large blossoms grow singly or in pairs, and the leaves standing on very short footstalks close to them, form a heavy background. The fruit is mild and mealy, and the core stony. The leaves are of a long lance-shaped form.

The Hawthorn (Cratǽgus oxycántha, *Plate VII., fig.* 9), is the one British representative of the last family in the great Rose tribe. Its snowy clusters of fragrant blossoms, sharp-lobed leaves, and red berries, do not need any description; they are familiar to us all from infancy. There is a town in the north of Yorkshire called after the fruit of the Hawthorn—"Hawes." The bush abounds in its neighbourhood. The Pink Hawthorn in plantations is a cultivated variety of this.

The Greeks regarded the Hawthorn as an emblem of hope; the Romans used torches of it in their marriage feasts and processions. A curious legend exists regarding a certain Hawthorn at Glastonbury. It says that Joseph of Arimathea came to preach in Britain, and landed upon the high grounds near Glastonbury, which was then an island; he fixed his staff in the ground, and fell asleep. When he awoke the staff was rooted and had borne leaves and flowers. He built a chapel near the spot, and Gilpin says that the Thorn flowered always at Christmas. But we have an historical association with the Hawthorn which has a better foundation than the Glastonbury fable. Miss Strickland tells us that when Richard III. was slain at Redmore, and his body plundered, a soldier hid the crown in a Hawthorn bush. It was soon found and carried to Lord Stanley, who placed it on the head of his son-in-law, saluting him by the title of Henry VII. In memory of this circumstance the House of Tudor assumed the device of a crown in a bush of fruited Hawthorn. This is doubtless the origin of the saying, "Cleave to the crown though it hang on a bush." May-day used to be celebrated throughout England, and branches of Hawthorn were hung over every doorway; this is still done at Athens. All poets have loved the Hawthorn, and celebrated its praises either alone or in connection with the floral May-day revels. Chaucer, in particular, depicts the favourite bush in a very pleasing manner:—

> "Amongst the many buds proclaiming May,
> Decking the fields in holiday array,
> Striving who shall surpass in bravourie,
> Marke the faire blooming of the Hawthorne tree;
> Who finely cloathed in a robe of white,
> Fills full the wanton eye with May's delight."

It is told of Bishop Latimer, that when he went to preach at a certain town, there was no one to hear him, for all the people had gone a-maying. One man came to him, and said, "Syr, this is a busy day with us, we cannot hear you. The parish has gone abroad to gather for Robin Hood. I pray you let them not." But the custom is little used now; here and there, in a remote village, the Maypole remains still, and children make garlands. At Hawkhurst this is the case. The Hawthorn is very useful for fences; skewers may be made of the large thorns, and a dingy yellow caterpillar, covered with hair, feeds on its leaves. The Black-veined white Butterfly comes from this caterpillar.

CHAPTER VIII.

ŒNOTHERÁCEÆ — HIPPURIDÁCEÆ — LYTHRÁCEÆ — CU-
CURBITÁCEÆ—FORMS OF FRUITS—PORTULÁCEÆ—
ILLECEBRÁCEÆ—CRASSULÁCEÆ—GROSSULARIÁCEÆ
—SAXIFRAGÁCEÆ—FORMS OF INFLORESCENCE.

> "In His Spirit God hath clothed the earth,
> And speaketh solemnly from tree and flower."
> *Beautiful Poetry.*

WE now come to several smaller and less important orders of British plants, each of which, however, has interest of one kind or another. The EVENING PRIMROSE order succeeds the Rosaceous in the subclass of Calyciflorals; its grand feature is the number Four. They have four sepals, four petals, and capsules of four valves.

The Isnardia has four petals (or none), four sepals, and four stamens.

The Enchanter's Nightshade has half the number—two petals, two sepals, two stamens. Our garden Fuchsias belong to this tribe; their fruit is an oval berry.

The true Willow-herbs derive their Latin name, Epilóbium, from two Greek words signifying *upon a pod*, because the flowers are placed at the top of a long seed-vessel. The seeds of this family are numerous, and tufted with down. On my first visit in Swaledale I was driving with some friends by the river side. On the opposite bank I saw some tall plants, covered, as I imagined, with delicate white flowers. In eager curiosity I questioned my host, hoping that a new plant was in store for me. I could imagine no white flower of that appearance, and became sanguine that it was something so rare that I had never seen or heard of it. My kind friend, anxious to give me pleasure, put the reins into my hand, and crossed the stream, partly by springing from stone to stone, and partly by wading. He returned with the desired boon, but, alas! instead of a new specimen, it was only my old shrubbery friend, the Rose-bay Willow-herb (Epilóbium angustifólium), half covered with its feathery seeds, which the long pods had discharged. This plant is tall and graceful, the large crimson flowers grow in a full spike, and the calyx and stem are tinged with red. It grows freely among rocks in watery places in Swaledale. Edward also describes it as forming a striking object in damp woods skirting the Hastings road.

Another large showy member of this family is the Hairy Willow-herb (E. hirsútum, *Plate VIII.*, *fig.* 1). Its flower is paler, and its foliage more conspicuous, than that of the Rose-bay or Narrow-leaved species. We used to call the plant "Apple-pie," or "Codlins and Cream," when we were children, on account of its smell.

The Small-flowered Willow-herb (E. parviflórum), grows on a clay bank in Birkpark wood, in Swaledale; it has downy foliage, a simple stem, and pale lilac flowers.

The Smooth Willow-herb (E. montánum), is a very common species; its pale flowers and broad-shouldered smooth leaves appear on waste ground very frequently. All these species have the stigma divided into four.

The Pale Willow-herb (E. róseum), is an inhabitant of marshes; I have not found it. This and the remaining species have undivided stigmas.

The Square-stalked Willow-herb I found in a watery lane high in Swaledale; its chief peculiarity is expressed in its name.

The Marsh Willow-herb (E. palústre), has narrow, slightly-toothed leaves, and a round stem. It is altogether a larger plant than any, except the two first Willow-herbs.

The Chickweed-leaved Willow-herb (E. alsinifólium), has a creeping root, and squarish stalk. I have no specimen of it.

The least of all the Willow-herbs is the Alpine. It is only

three inches high, and not branched; it bears only two flowers, and its leaves are half as broad as long (E. alpínum). It flourishes on a peat moor called "Summer Lodge-bank." The young shoots of some of the Willow-herbs are cooked and eaten, and in Kamschatka an intoxicating liquor is made from them. Although the number Four is the leading characteristic of this group, the Willow-herb family have eight stamens.

There is but one British representative of the Evening Primrose family (Œnothéra biénnis, *Plate VIII., fig.* 2). The plant is tall, thick, and herbaceous, and bears large, beautiful, lemon-coloured flowers. My specimen grew on waste ground about Clevedon, flowering in July. This plant belonged originally to America, but it was introduced into Europe by the French, who cultivate the roots for food. The petals open in a very curious manner. The calyx has little hooks on the upper part, by which it holds together when in bud. The sepals open first at the lower part, and then the corolla bursts asunder the hooks with a loud noise. The flower remains open all the night, unfolding about six in the evening, and closing again as the morning advances. There are many handsome members of this family cultivated in gardens, of every shade of yellow, and some snow white. The poet thus notices the Evening Primrose:—

> "When once the sun sinks in the west,
> And dew-drops pearl the evening's breast;
> Almost as pale as moonbeams are,
> Or its companionable star,
> The Evening Primrose opes anew
> Its dainty blossoms to the dew."

The Marsh Isnárdia (I. palústris), is a very scarce, floating pond-plant. The ovate leaves are placed opposite to each other, and are bright green and succulent; the small green flowers are solitary. I have seen drawings of the plant, but have found no specimen, so we must pass over this family of the Willow-herb group.

Next to it comes the Enchanter's Nightshade (Circæa lutetiána). This is a pretty plant, with bright green lance-shaped leaves, and slender spikes of white or pinkish flowers. The sepals, petals, and stamens are only two.

There is an Alpine species, of lower growth, and less branched; its leaves are heart-shaped. The Common Enchanter's Nightshade I have found frequently in woods. Although called after an enchanter, the plant has no charm except quiet beauty; and it is supposed that it owes its name to the darkness and gloom of the places where it grows.

The MARE'S-TAIL order are all water plants. Fanny reaped a rich harvest of them one day at Clevedon. There is a large tract of lowland extending two or three miles inland called Kenmore; this ground is intersected with ditches. In one of these ditches the common Mare's-tail (Hippúris vulgáris), grows; it roots in the muddy bottom, and its long stem soon appears above the water; quickly one whorl of awl-shaped leaves after another arises, in the axils of which grow tiny crimson flowers, each boasting one stamen and one stigma, and no petal.

Her next trophy was the Spiked Water Milfoil (Myriophýllum spicátum, *fig.* 5). It has whorls of four grass-shaped, pinnate leaves, and spikes of small pink flowers, with four petals, four styles, and eight stamens. There was a tangled mass of this in one broad ditch; most part of the plants under water, and only the pink spikes rising into the air.

The Whorled Water Milfoil has its flowers in a whorl like the Mare's-tail, but we have not found a specimen.

In a clearer ditch she found the Vernal and Autumnal Water Starworts. The former species (Callítriche vérna, *fig.* 6), has broadish leaves, the upper four or six floating on the surface, and looking like a green star; while the tiny flowers, with their one style and stamen, proceed from their axils.

The Autumnal Starwort begins to flower in June; its leaves are more slender, and it grows more under water.

There is another species, with keeled seeds, called the

Pedunculated Starwort; it grows only in Sussex, and I have no specimen. These are all the members of the Mare's-tail group.

The Hornwort group comes next. Fanny found one member of its only genus in the same fruitful ditch on Kenmoor. The common Hornwort (Ceratophýllum demérsum), has long stems thickly beset with compound bristly leaves, at the foot of which are whorls of tiny flowers. These have many stamens, but they are borne on one petalless flower, and the sessile stigma on another. The fruit is crimson, with a horn at the end.

The only other British species, the Unarmed Hornwort (C. submérsum), is distinguished by having no horns on its seeds.

Edward has specimens of some of the LOOSESTRIFE order. In this tribe the calyx is tubular; cloven into twelve teeth. There are six petals, and the stamens are either six or twelve.

The Spiked Purple Loosestrife (Lýthrum salicária, *Plate VIII., fig.* 3), is a very handsome plant, growing four feet high, with long lance-shaped leaves placed opposite to each other, squarish stems, and spikes of crimson flowers set on in whorls. It grows by streams and in watery places, flowering in July and August. He has gathered it on the banks of the Avon, in Warwickshire, and in watery places about Hawkhurst, Kent.

The Hyssop-leaved Purple Loosestrife is a smaller plant, inhabiting much the same situations as the Willow-leaved species, but its flowers are less brilliant in colour, as well as smaller; and it is very rare. We have no specimen.

The Water Purslane (Péplis pórtula), is a very insignificant plant. It grows from one to two inches high, with red stems, and broad, glossy, opposite leaves. The tiny flowers with their bell-shaped calices and six minute petals, are seated in the axils of the leaves. The plant grows in shallow water; when the water dries up its stem and leaves become entirely red. My specimens were gathered in and near a pond at Hawkhurst;

those away from the water were red, the others green. These are all the British members of the Loosestrife tribe.

The Henna belongs to this family; its uses are, according to Hooker, of a very remote date. The Moors, Arabs, and Turks cultivate it at the present day, and use it to dye their hair and nails, and also the backs, manes, and hoofs of their horses. The women abstain from it at the death of their parents and husbands. The leaves are gathered in the spring, then dried in the air, and reduced to powder, and applied in the form of paste to the parts they wish to tinge. The foliage is also used to cure recent wounds and abscesses. Its flowers have a strong penetrating odour; by distillation an extract is obtained which is used in the baths, and as a perfume in visits and religious ceremonies. It was no doubt on account of their odour that Henna flowers were scattered by the Hebrews in the apartments of a bride, and for the same reason the Egyptians keep it in their rooms. A considerable trade is carried on in Henna leaves, which yields a large revenue to Egypt. Moore sings of this plant :—

> " Thus some bring leaves of Henna to imbue
> The finger ends of a bright roseate hue;
> So bright that in the mirror's depth they seem
> L'ke tips of coral branches in the stream."

The GOURD order comes next in succession. The characteristics of this tribe are extremely contradictory; there is at once a sweet and nutritious, and a bitter and poisonous element in each member, and the plant is poisonous or wholesome according as the one or other quality prevails. The Melon, Vegetable Marrow, and Squash are examples of the sweet and nutritious part of the tribe; the Bottle Gourd, Colocynth, and Bryony, of the bitter and poisonous portion. Mr. Johns asserts that the "wild vine," off which one of the sons of the prophets gathered "wild Gourds a lapful," in the time of Elisha, was one of this tribe. We all know how very quickly these plants grow, and the example of Jonah's Gourd at once recurs to the mind. It conveys a touching lesson regarding all earthly blessings: had the impatient Prophet been exceedingly thankful to God for the Gourd, instead of "exceeding glad" of it, his mind would have become softened and subdued, and he would not have needed to be taught pity for his fellows by the loss of his own shelter. When God "prepares a worm" to smite our Gourds, it is because His tender mercies and abundant blessings have failed in teaching us love to Him, and to our brother man. This we must learn somehow, and if the withering of our Gourd does not effect it a "vehement east wind" must come, and the sun must smite upon our head. The fruits of the nutritious members of the Gourd tribe are well known to us, either by hearing or by personal experience. But even from the poisonous kinds some good is elicited; medical art discovered in the Colocynth a valuable medicine, whilst the homœopathic doses of our own Bryony work wonders in allaying spasmodic coughs. "This also cometh of the Lord of Hosts, who is wonderful in counsel, and excellent in working."

Our own Bryony (Bryónia dioíca, *Plate VIII.*, *fig.* 4), is a graceful climbing plant with broad leaves, having five pointed lobes, and rough both above and below. It has clusters of pale green flowers, beautifully veined with darker green. The stamens are in one flower, and the style in another. Its long clusters of round scarlet berries form a beautiful object in autumn. I have gathered specimens both in the neighbourhood of Ripon and in Wensleydale, Yorkshire.

The one British member of the PURSLANE tribe (Móntia fontána), has also fallen in my way. It is a small pale green plant, with spare clusters of tiny white flowers, boasting five petals, and four stamens. It grows in watery places—mine is from Swaledale, and has the name of "Blinks." There are exceedingly handsome members of this family growing wild at the Cape of Good Hope, and they are much sought after in our gardens and greenhouses. Blue and scarlet are their prevailing colours.

FORMS OF FRUITS.

It is time now that we should give attention to the different forms of fruit, and gain a clear understanding of the terms applied to them. An Apple cut in two exhibits a five-celled core enclosing the seeds, the whole surrounded by the swollen juicy flesh of the calyx. This is the type of the *pome* (*fig.* A).

In the *drupe* we have one seed contained in a woody case, and surrounded by a fleshy pulp. The sketch of a half Peach (*fig.* B), is a good example. Another form of arrangement of

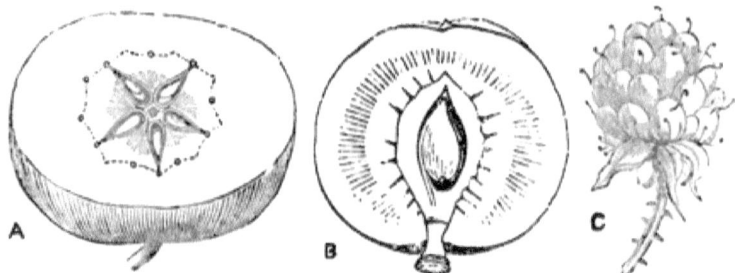

drupes is in the Bramble, where a number of them are collected together around a receptacle (*fig.* C); the Strawberry fruit is also fixed upon a receptacle, but the seeds are naked and embedded in pulp (*fig.* D).

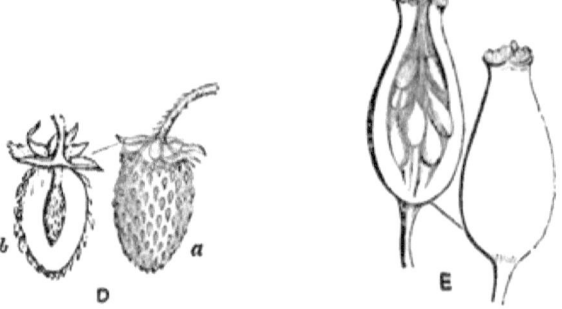

The Rose fruit or hip is a collection of hairy, nut-like seeds, enveloped in the enlarged and fleshy tube of the calyx, and carrying the sepals as a crown on its summit (*fig.* E). We have the *follicle*, the kind of vessel in which the Pæony carries its seed, a kind of pod with one lateral opening (*fig.* F); the *capsule*, exhibited in the Poppy tribe, and in many others (*fig.* G), the Lychnis family, for instance, where it is enveloped in the permanent calyx. Then there is the *pod*, which characterises one great division of the cruciform plants, as the Wallflower, Cuckoo-flower, &c. (*fig.* I), and the

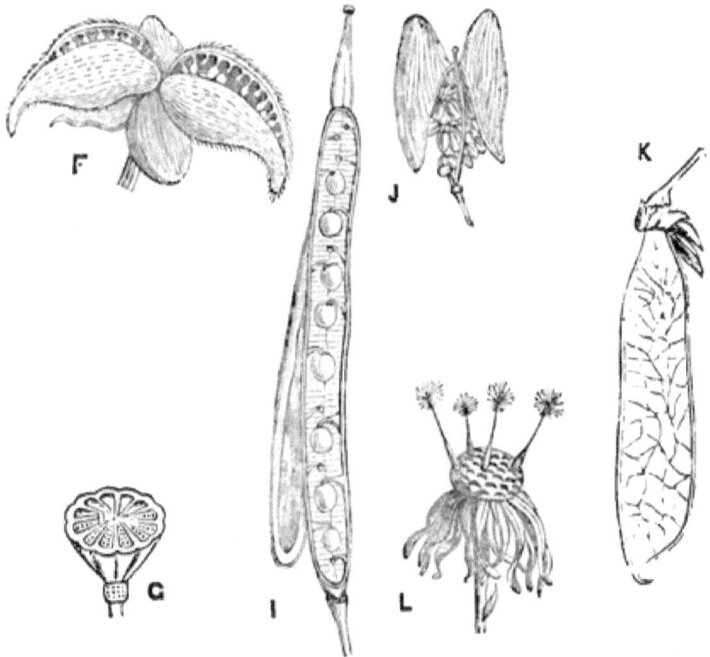

pouch, the seed-vessel of the other class of cruciform plants—the Shepherd's Purse, Pepperwort, and Scurvy-grasses, for instance (*fig.* J). The *legume* is the mark of a very important tribe, both for number and utility; our edible Peas and Beans form familiar examples (*fig.* K). The seeds of the Composite tribe, endowed with a *plume*, by means of which they are scattered, are very interesting. The "clocks" of the Dandelion are familiar to us all (*fig.* L).

The Hazel-nut is a fruit of a bony structure enveloped in a tubular laciniated coat which botanists call an involucre (*fig.* M). One other variety of fruit is here—the *winged-seed-nut* (*fig.* N), which characterises the fruit of the Sycamore

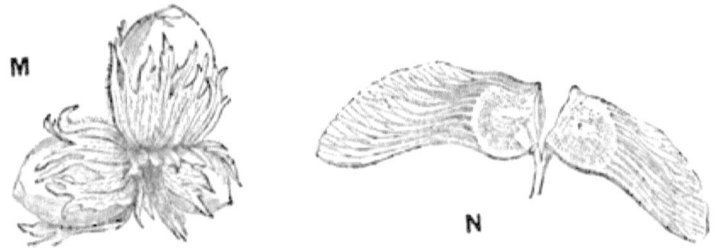

family. This seed-bearing quality in plants is a proof of their pertaining to the third day's creation. " And God said, Let the earth bring forth grass, the herb yielding seed, and the fruit tree yielding fruit after his kind; and it was so." And it is further the title by which the great Creator made the vegetable world over to His own crowning work—man. "Behold, I have given you every herb bearing seed, and every tree in the which is the fruit of a tree yielding seed." The one exception to this rule—the one tree whose fruit was withheld, was, as you know, the occasion of the fall of our first parents. But that error being atoned, our God has promised a new Paradise, where the Tree of Life will flourish, bearing twelve manner of fruits; not yearly, but monthly. We know not what blessed quality that fruit may have, but we are told that the *leaves* of the tree shall be " for the healing of the nations."

To return to the plants before us: the KNOT-GRASS family comes next in order. Fanny hoped to have found the Strapwort near the Loe Pool, in Cornwall, where Mr. Johns describes it as growing, but she sought it in vain. It takes its name from the strap-shape of the leaves; and its small white flowers growing in tufts have five petals, five stamens, and three stigmas. Its style of growth is prostrate. She was more fortunate in her search for the Rupturewort, to the locality of which she was directed by the same writer. In the famous excursion along the cliffs from Kynance to the Lizard Point in which she found the Bloody Cranesbill, Common Dropwort, English Scurvy-grass, and other treasures, they passed along the cliffs at the foot of the Soap Rock—a mass of fine soft stone of dazzling whiteness, contrasting strangely with the rugged serpentine cliffs so dark in their hue, which border the coast on either side of it: growing amongst the débris of this Soap Rock was a plot of the tiny prostrate Rupturewort. The flowers are of the same colour as the rest of the plant, and have five petals, five stamens, and one stigma. There are two kinds of Rupturewort, one smooth and the other hairy, and Sir J. E. Smith considers them different species. Fanny's specimen is hairy (Herniária hirsútum); it is the more scarce of the two.

The Smooth one (H. glábra), I have seen in the botanic garden at Edinburgh.

In a swampy meadow near the Loe Pool, Fanny found a group of dainty plants: the Pale Butterwort, Ivy-leaved Bellflower, and Whorled Knot-grass. This last is the only one of them with which we have to do at present. It has a slender stem three or four inches high, along which oval glaucous leaves are planted in pairs at short intervals. A whorl of pinkish flowers graces each pair of leaves nestling in the axils. Each blossom has five coloured sepals, no corolla, five stamens, and one stigma; the stem is partly recumbent (Illécebrum verticillátum). She failed in finding the Four-leaved Allseed, the lower leaves of which grow in fours. It has five petals and five sepals, three stamens and three stigmas. It is a tiny plant, with a cluster of flowers in a form called a *panicle*.

One family is still wanting to complete the Knot-grass tribe, and I am happy to have a specimen of one of its members.

The Annual Knawell (Scleránthus ánnuus), first met my eye

in a certain sandy lane in Cheshire, to which I have already referred more than once. It is a light, whitish-green plant, growing partly prostrate; it has no corolla, and its five-cleft calyx has a thin white margin. The flowers grow in clusters both at the end and in the forks of the stems; the leaves are small, narrow, and pointed. I have no specimen of the Perennial Knawell, but it very nearly resembles this, being distinguished by its more entirely prostrate stems and blunt leaves. It frequents the sandy fields of Norfolk and Suffolk.

The next tribe to that of the Knot-grass is the STONECROP tribe. Our British Stonecrops are very humble members of the race. They inhabit every zone, but the most beautiful species are found in the tropics. The scorched cliffs of the Canaries are adorned with these plants, and the sterile plains of the Cape of Good Hope are relieved by their presence. I remember while travelling in Switzerland, a lady of our party directed our attention to a magnificent plant of this tribe suspended from some rocks in a steep ravine beneath us. Its very large panicles of white flowers had the effect of Ostrich feathers. Our friend admired the plant enthusiastically, and expressed a strong desire for it to place among her collection of Swiss plants; a moment after she regretted having done so, for her young husband was climbing down the rocks, which overhung a foaming mountain torrent, to procure it. However, he returned in safety with his prize, which he placed in the hat of his wife. No plume of feathers ever had a more graceful effect, and she wore it under the blazing sun during the whole day's excursion, without one of the delicate petals flagging. In fact, the principal interest attaching to these plants arises from their needing so little apparent sustenance from the soil. Where other plants pine and die these flourish; the slightest hold upon the rock or sand sufficing for them. True and beautiful is the lesson that the poet puts into their mouth; we would fain sit as loosely to the earth as they, but have a more real and secure attraction above. They teach us not to set our affections on things on earth, and the Spirit that breathes through all creation, as well as through all inspiration, adds, "Set your affections on things above," for "your life is hid with Christ in God."

> "There from his rocky pulpit I heard cry
> The Stonecrop: 'See how loose to earth I grow,
> And draw my juicy nurture from the sky;
> So draw not thou, fond man, thy root too low.
> But loosely clinging here,
> From God's supernal sphere
> Draw life's unearthly food—catch heaven's undying glow.'"
> REV. R. W. EVANS.

The Tillæa is the first family of the Stonecrop tribe. Common as Mr. Johns declares the Mossy Tillæa to be, it has never come in my way. A weedy plant, resembling moss, with procumbent stems and sessile flowers, mostly three-cleft. I know it only from Smith and Sowerby's beautiful illustration.

The Common Navelwort (Cotylédon umbilícus), is the chief if not the sole member of its family. With its shield-shaped leaves, and upright racemes of waxy, bell-shaped flowers, it is a familiar ornament of old walls and hedgebanks in Devonshire. I have also found it abundantly about Ludlow, and on rocks near Ross, Herefordshire. My specimen was gathered near Tiverton in July, and during the same month other specimens were sent me from Linton.

Sir J. E. Smith mentions a second species—the Yellow Navelwort; but he says it is introduced as British by error.

The third family of the Stonecrops is the Houseleek (Sempervívum tectórum, *Plate VIII., fig.* 5), and it has only one British member: a handsome, starry flower, with twelve petals, and twelve or more stamens, large, and of a bright rose. The thick, overlapping leaves are of a blue green tinge, rough at the edges, and bordered with red. The blossoms grow in a kind of straggling cyme, or three-branched raceme. It is a familiar ornament of roofs and walls; my piece was gathered at Birk Park, in Swaledale. The leaves of the Houseleek have cooling properties, and are much used in country places to lay upon wounds. From a foreign species an excellent tannin is prepared.

Now for the Stonecrops proper. Their Latin name (Sédum), is from a word signifying to *sit*, on account, I suppose, of their low growth. They have five petals, five sepals, five stigmas, and ten stamens. The leaves are thick, and mostly glaucous, and the flowers small.

The tallest, but not the prettiest of the family, is the Orpine (S. teléphium). It grows in woody places, and in hedgerows, attaining the height of two-and-a-half feet. The leaves are egg-shaped and serrated, and the dingy purple flowers compressed into a close cluster. The plant grows freely about Hawkhurst, and I have specimens from Cheshire. August is its flowering time. It is the only Stonecrop with flat leaves.

The Thick-leaved Stonecrop (S. dasyphýllum), grows on walls about Clevedon. It is a small plant, with weak stems and almost round leaves, which are tinged with red. The little flowers are white.

The White English Stonecrop (S. ánglicum), much resembles the last species, but is of a stronger habit, and its white petals are prettily spotted with crimson. It grows in quantities on the Looe cliffs, and very ornamental it is. Both these specimens were gathered in June.

Another member of the family, the Rock Stonecrop (S. rupéstre), grows on those same cliffs. It has a handsome cyme of yellow flowers, and grows at least a foot high; its leaves are awl-shaped and spurred.

My contribution to this family would have been but small, had it not been for Mr. Ward's kindness. The walls and rocks in Swaledale furnish abundance of the Biting Stonecrop or Golden Moss (S. ácre, *Plate VIII.*, *fig.* 6). Its three-branched cyme of yellow flowers is a familiar object, and the rocks owe much of their beauty to it. The leaves are thick and spurred; they are placed alternately upon the stem.

The Insipid Stonecrop (S. sexanguláre), was given to me by Mr. Ward. It is rather a taller plant than the last, though so much resembling it, and the leaves are placed in six rows. At the same time he gave me the White Stonecrop, the flowers of which form a smooth, colourless panicle. The stalks are purple, and the leaves drop off very soon. The Welsh Stonecrop is also his gift. It bears a cyme of bright, yellow flowers. It is distinguished from the Rock Stonecrop by the brightness of its leaves, whilst the Crooked Stonecrop is marked by its leaves being turned back.

The distinguishing characteristics of the Hairy and Glaucous Stonecrops are expressed in their names. I have not been able to find specimens of either.

The Rhodiola, or Rose-root, is the only species of the last family of Stonecrops. It is a stout plant, with a woody root, the scent of which resembles that of the Rose. Its broad, thick, pointed leaves are often folded over one another, and the cyme is composed of minute dirty yellow flowers. I know many places in the Scotch hill country where it is found, but I am obliged to be content with a garden specimen.

The brilliant Mesembryanthemums of our gardens, and the gorgeous Cactuses, intervene between the Stonecrop tribe and that succeeding it in British botany—the Gooseberry.

This is an important tribe to us for the service we have from the fruits. The Wild Gooseberry (Ribes grossulária), is found growing freely in woods in the North of Yorkshire. Beautiful fruit, fully ripe, was found wild in Wensleydale last year. This shrub is very thorny, bears its flowers singly or in twos, and has a hairy berry. The boys of the district seldom allow these berries to ripen. Our garden Gooseberries are varieties of this, and like it flourish well in a cold climate. Murray states in his "Encyclopædia of Geography," that they grow in remarkable luxuriance in vallies in Aberdeenshire, 900 feet above the level of the sea. He cites Bishop Forbes, of Aberdeen, who died in

1635, as remarking that this fruit becomes more delicious as it approaches the upper regions of its cultivation.

The Red Currant (Ríbes rúbrum, *Plate VIII.*, *fig.* 7), has no thorns; it has drooping racemes of yellowish-green blossoms. It grows to the height of seven feet on the banks of the Swale, both below Richmond, and high up into Swaledale; many of the places where it flourishes are very far from any garden, so that I feel assured of my specimens being wild.

The Rock Currant (Ríbes petrǽum), carries its clusters erect when in bloom, but they droop as the fruit becomes heavy. This shrub abounds in rocky woods about Richmond. Its leaves are downy underneath.

The Mountain Currant (R. alpínum), we long sought in vain, but we found it at last in a grove, a part of the property of Thiernswood, in Swaledale. The situation is very high, and the wood is extremely rocky. This Currant has smooth, shiny leaves, and carries its raceme erect both in flower and fruit.

The Black Currant (R. nígrum), the last of the Gooseberry tribe, used to grow wild in woods near Reeth, but we spent a whole day seeking it last year, and yet did not succeed in finding it. Edward has the advantage of me, for he had the pleasure of taking it prisoner in Kent last year. The bush was growing in a very shady place, on the bank of one of the slow, muddy streams. It had to thank its position for its being in flower in June. The strong odour of the leaves, more robust habit of the plant, and brown tinge in the spare flower-cluster, distinguish this from any of the other species. I fear the associations with the Gooseberry tribe are entirely edible. The name brings at once to one's mind pies and puddings, jams and jellies.

Leaving the Gooseberries we now turn our attention to the SAXIFRAGE order. This tribe contains two families: that of the Saxifrage, and of the Golden Saxifrage. These have five sepals and five petals, ten stamens, and two styles.

They are all plants of humble growth, frequenting temperate climates.

The White Meadow Saxifrage (Saxifraga granuláta), is our most common species. The intense whiteness of its somewhat bell-shaped flowers, which grow several together, make it a very attractive plant. Its leaves are round or kidney-shaped, and bluntly lobed. The root has fleshy grains or knobs in it, hence its Latin name.

At the time when the doctrine of "signatures" was in vogue these knobby roots were supposed to be a cure for swellings and all excrescences growing upon the human frame. This theory tried to prove that any resemblance in a plant showed that it was an antidote to the disease of the part of the human body to which it bore the likeness. Thus the spotted leaf had a fancied likeness to the human lungs, and was therefore declared to be a cure for consumption; and the granules of the Saxifrage formed a remedy for hard excrescences. It was a superstitious, pagan-like foreshadowing of the homœpathic rule, "Like cures like," only it stopped short in surface-resemblance, instead of taking its stand on similitude of nature. But all this has been exploded long ago, and our little Meadow Saxifrage has now no useful service to perform.

The little Rue-leaved Saxifrage is pretty frequent, adorning old walls with its tiny blossoms and leaves, more or less coloured with red. I have gathered it on the ruins of Easby Abbey, and on old walls about Richmond.

The Mossy Saxifrage, or Eve's Cushion (S. hypnoídes, *Plate VIII., fig.* 8), grows in the rocky parts of Swaledale, both on the brink of the river and high upon the hills. It forms large mossy plats, its leaves are much divided, and their segments narrow; and its clusters of bloom, though of a less clear white and smaller, are scarcely less pretty than those of the Meadow Saxifrage.

The Purple Saxifrage grows freely among rocks about the

English Lakes; its flowers grow singly on the stems, and the closely-packed leaves are placed opposite to one another.

The Yellow Marsh Saxifrage (S. hírculus), is found on Knutsford Moor, in Cheshire, not many miles from my favourite Rudd Heath. The erect stem is clothed with narrow, alternate leaves. It reaches the height of eight inches, and the large yellow blossom is beautifully speckled with red.

The London Pride Saxifrage (S. umbrósa), is an Irish plant, but I gathered it, apparently wild, in the Aske woods, near Richmond. It is too common an ornament in gardens to need any description. A species with large white flowers spotted with yellow, very slight in its form, called the Starry Saxifrage, was given to me by Mr. Ward. It frequents the neighbourhood of alpine rills.

The Yellow Mountain Saxifrage (S. aizoídes), is a gay little plant; its blossoms spotted with red. It greatly resembles the Golden Moss, but is darker in hue, more of an orange colour, and also larger. I found it abundantly in Switzerland and in the Scotch Highlands, and my specimen is from thence.

There are a Kidney-leaved and a Hairy species, both resembling the London Pride; and, like it, they are inhabitants of Ireland.

The Drooping, Alpine, and Tufted Saxifrages are white species peculiar to mountain country; the Moss-like and Dwarf ones are yellow, and inhabit similar localities.

The Long-stalked, Broad-petalled, and Web-footed Saxifrages, have brilliant white blossoms, and are found in Scotland and Wales. We have no specimens of any of these.

I have both the Golden Saxifrages; they have no corolla, but the calyx and stamens are of a brilliant gold colour. The two species are distinguished from one another by the position of their leaves, those in the one being alternate, and in the other opposite. Both bear a flat cyme or corymb of yellow flowers, and their leaves are kidney-shaped and notched.

The Alternate-leaved species is common, growing abundantly in wet, shady places in high and low lands.

The Opposite-leaved (C. oppositifolium), species is much more rare, but it grows, mingled with the Alternate-leaved on Billy Bank, near Richmond. These complete the Saxifrage group.

FORMS OF INFLORESCENCE.

In describing these recent orders, I have used various terms for different kinds of clusters, which I will now proceed to explain.

A *panicle* is a loose cluster of flowers on branched footstalks of various lengths, springing from the main stem at different places. Many of the Grasses flower in panicles (*fig.* A). The other general forms of clusters are:—the *spike*, where the flowers grow up the stem without having any footstalks; the

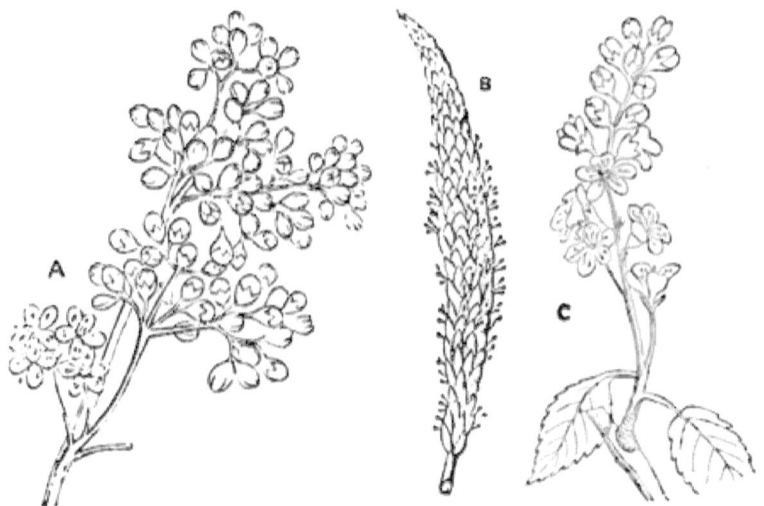

Beet, Broom, Rape, and Ladies' Tresses, are examples of this (*fig.* B); the *raceme*, where the flowers are placed on simple footstalks, as in Currant (*fig.* C), Bird Cherry, and

Laburnum; the *cyme*, where all the footstalks spring from one spot, branching in various directions, and forming a round or flattish surface, as in Cornel and Gueldres Rose (*fig.* D); the

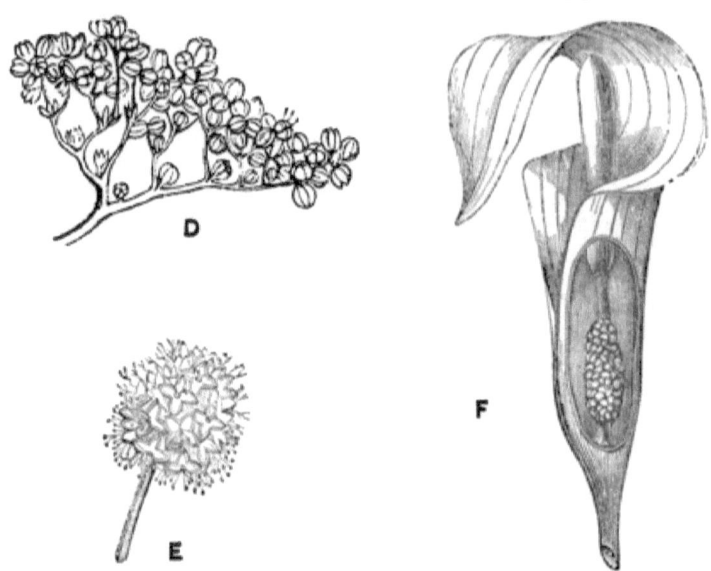

head, where the seated flowers are placed in a compact cluster, as in Clover, Burnet, and the Compound tribe (*fig.* E); and the *spadix and spathe*, where the flowers proceed from a sheath, as in the Wild Arum, Narcissus, and Snowflake (*fig.* F).

CHAPTER IX.

UMBELLÍFERÆ—ARALIÁCEÆ—CORNÁCEÆ.

> "To every Form of Being is assigned
> An *active* principle; howe'er removed
> From sense and observation, it subsists
> In all things, in all natures; in the stars
> Of azure heaven, the unending clouds,
> In flower and tree, in every pebbly stone
> That paves the brooks, the stationary rocks,
> The moving waters, and the invisible air.
> Whate'er exists hath properties that spread
> Beyond itself, communicating good,
> A simple blessing, or with evil mixed;
> Spirit that knows no insulated spot,
> No chasm, no solitude; from link to link
> It circulates, the Soul of all the World."
>
> WORDSWORTH.

THE large order upon which we are now entering, that of the UMBELLIFERS, abounds in interest. The various members of the families that compose it have minute flowers growing singly on very small stems, all which stems are collected in a cluster and united to the parent stem at the same point; the cluster thus formed is called an umbel, because the stems radiate in a circle like the ribs of an umbrella. In many of the plants the umbels are compound—that is, the main cluster is formed of a number of small ones. Each little flower has five petals, a calyx formed either of five sepals or a mere ridge, five stamens, and ovary of two cells. The fruit consists of two carpels united at the face, and with raised stripes along their sides. The prevailing colour of the flowers is white, but some are yellow, some tinged with lilac, and a few blue.

At the juncture of the flower-stalks with the main stem there are generally small leaves or *bracts* placed in a whorl round the stem; these are called, collectively, an *involucre*.

The order is very extensive, numbering 1500 species. The only reliable mark of distinction is the seed, and we shall examine them with reference to this, also describing them as their appearance strikes the more superficial observer. This order has members possessing the most varied qualities. Some are rank poison, some merely unwholesome, some aromatic, some edible. As a general rule, those growing "high and dry" are wholesome, while those frequenting water are more or less poisonous.

Assafœtida is a plant of this tribe, indigenous in Persia. Gum galbanum is extracted from another. Our wild Hemlock is much used in medicine; as are also Coriander, Dill, and Anise. Carrots, Parsnips, and Earthnuts have edible roots; Celery, when cultivated, has wholesome stems; Parsley and Fennel are esteemed for their leaves, and Samphire makes a good pickle.

The Coriander has a round fruit dividing into two hemispheres; the lower leaves are pinnate; the upper, thrice ternate and thread-shaped. It frequents fields and dunghills, and is not truly wild. The seed is aromatic, and used in making comfits. The manna of the Israelites was likened to it—"the manna was as Coriander seed." (Numbers xi. 7.)

Galbanum, an ingredient in the incense of the sanctuary, was obtained from the Galbanum officinale, an umbelliferous plant of Syria.

The Shepherd's Needle (Scándix pecten-Véneris, *Plate IX.*, *fig.* 6), is familiar to us all with its tall beaked fruit, like a bunch of large needles. The umbels are simple and inconspicuous; flowers white, and leaves very narrow and repeatedly pinnate. We have all specimens of this.

The Beaked Parsley (Anthríscus vulgáris), has a very short beak, and its fruit is rough with bristles. It has no calyx, the stem is swelled under each joint, and the doubly pinnate leaves are of a bright green. It frequents waste places.

The Wild Chervil (Chærophýllum sylvéstre), is a plant easy to distinguish; its full green leaves are very large, composed of numerous small leaflets. The umbels are large and widely spread, the flowers white. The scent of the plant is somewhat aromatic.

The Garden Chervil (C. satívum), has the umbels from the sides of the stem. As in the two last families, the fruit is beaked, and there is no calyx.

The third group, according to Sir J. E. Smith, comprises those genera whose fruit is solid and prickly, without a beak. Our British families of this kind are the Sea Holly, the Carrot, the Sanicle, and the Bur and Hedge Parsley.

The Sea Holly is one of Fanny's trophies; it grew on the bar of sand separating the Loe Pool from the ocean. Its umbel is simple, and the flowers blue; the leaves are lobed, acute and thorny; the colour a very blue or glaucous green, and the veins white (Erýngium marítimum, *Plate IX., fig.* 3). There is a Field Eryngo, with lilac flowers and bright green foliage, but it is doubtfully indigenous. The Sea Holly stems are preserved and make a pleasant candy.

The Sanicle (Sanícula europǽa, *Plate IX., fig.* 2), I have gathered in woods in Somersetshire, Kent, and Yorkshire in June, and I have seen the leaves, later in the summer, among the herbage of woods in other counties. It is a handsome plant, with dark, glossy foliage, the leaves round and cut into deep lobes; and the umbels carry their flowers so close to the stem that they seemed quite globular; the long, white stamens give them a light appearance.

We have the Wild Carrot (Dáucus caróta, *Plate IX., fig.* 8), both from Kent and Cornwall. The umbel is large, but the stalks in the centre are shorter than those round the margin, so that the whole collection of small umbels seem to form a cup, in the centre of which is one crimson flower. The leaves are much divided, and the segments pointed. The root is tap-shaped, doubtless the original of our garden Carrot, the utility of which is patent to all.

The Sea Carrot (D. marítima), Fanny describes as growing

upon the Cornish cliffs, from Talland to West Looe. Its leaves are thicker, and cut into broader segments than in the common species, and it lacks the characteristic crimson flower. In seed the central umbels rise higher than those at the edge, so as to give a hemispherical form to the main umbel.

The Knotted Hedge Parsley (Torílis nodósa), was growing beside the Sea Carrot; its umbels are simple, very minute, and grow from the axils of the leaves. Its seeds are covered with knots, hence its name.

The Spreading Hedge Parsley (T. infésta), abounds as a weed in the corn fields of Kent. It is a low-growing, shrubby plant, with whitish-green foliage, twice pinnate leaves, and small umbels of white flowers.

The Upright Hedge Parsley is very common; we have specimens from Yorkshire, Durham, Cheshire, and Warwickshire. It is an elegant plant, often growing as high as three feet. It is prettily tinged with purple at the tips of the fruit, leaves, and on the stem. Its spotted stem made me at first mistake it for Hemlock, but the roughness of the whole plant, and the furrows on the stem, soon undeceived me. The umbels nod when in bud; the flowers are white.

The Bur Parsley, with its bristly fruit, I have not found.

The next group has fruit without either wings or bristles, but it is ribbed instead.

The Sweet Cicely (Mýrrhis odoráta), is of this family. Its scent announces it at the distance of many yards, strongly resembling that of Aniseed. The foliage is abundant, and of a light green; the plant tall and stout, and the white umbels large. Boys feed their rabbits with it.

The Rough Cicely has its seeds but slightly furrowed, and the rough stem is swelled under each joint. The leaves are thrice pinnate, and the flowers white. There is a species with orange fruit peculiar to Scotland. I have no specimen of it.

The Earthnut, or Pignut (Búnium flexuósum), as it is sometimes called, I have found in every part of England and Scotland that I have visited. Pigs grub for the nut-shaped root, and boys are equally alive to its charms. There was a paper in the *Gardeners' Chronicle* some years ago recommending the cultivation of this plant, because of the edible properties of the roots, which, when roasted, are as good as Chestnuts. The leaves resemble those of the Fennel, and the umbels are tolerably large. It is the earliest of Umbellifers in flowering, opening its white blossoms early in May.

The Water Dropworts are a remarkable family, some unwholesome, some actually poisonous.

The Tubular Water Dropwort (Œnánthe fistulósa), is really handsome, with thick hollow stems, and pinnate tubular leaves. It is a tall plant, upright, and with spare foliage of a glaucous hue. The general umbel has three or four branches, but the partial umbels are large, so that the cluster is a good size. The flowers are prettily tinged with pink.

The Parsley Water Dropwort is in every way a smaller, slighter plant, and its umbel has more rays. The two grow together in ditches, on Kenn Moor, near Clevedon, flowering in July.

The most harmful member of the family is the Hemlock Water Dropwort (Œ. crocáta), and it grows luxuriantly in ditches and watery places in the beautiful lanes and groves of Kent. The Hemlock Water Dropwort is the stateliest plant of the tribe, as far as my experience goes. Its foliage is formed of large leaves, combining numerous wedge-shaped leaflets; it is of a bright, glossy green, and surmounted with the large white hemispherical umbels, it cannot fail to attract attention. Its root is composed of fleshy knobs, each somewhat resembling the root of a Parsnip, and the whole plant abounds in a bad-smelling yellow juice. The farmers told Edward that it had killed horses, and seriously disordered cows. The principal danger, they said, is in the early spring, when the cattle are first turned into the pastures. They are then so new-fangled with every green thing that they eat the Hemlock Dropwort eagerly. Its shoots being then young are milder in their flavour than afterwards. Sad stories are on record of persons eating the roots, and illness or death being the result. Happily it is easy to recognise the plant when once you have been introduced to it.

Not entirely guiltless of poisonous quality is the Fine-leaved Water Dropwort (Œ. phellándrium). It grows as tall as the noble Hemlock Dropwort, but its umbels are not so large. Its foliage is cut into the finest possible wedge-shaped segments, and the plant has at once a stately and elegant appearance. Edward found it at Hawkhurst, growing up to his knees in the water of some fish-ponds.

The Sulphur Water Dropwort, with its thick stems and large fruit has not come in our way. Sir J. E. Smith speaks of it as rare.

We now come to that noted plant the Samphire (Crithmum maritimum, *Plate IX.*, *fig.* 7), once so much sought after even at the risk of life as a pickle for the table of the epicure. It is a fleshy, glaucous plant about a foot high, with much-divided leaves, and an umbel of greenish flowers. It always grows on rocks and cliffs above the reach of the highest tide. No shore is more favoured by this plant than that of Cornwall; it adorns the cliffs for miles, and flourishes in richest luxuriance all around Looe island. Dr. Murray in his "Physiology of Plants," gives a very interesting story connected with this plant, which I cannot repeat better than in his own words:—" During a storm, in November, 1821, a vessel was driven on shore near Beachy Head. It appears the entire crew were washed overboard, a few escaped to the rocks, apparently only to await a more lingering death. Each advancing wave narrowed the circumference of the little circle. They had already retreated to the highest point, and while debating whether to cast themselves into the waves, and trust the desperate event, or remain to be swallowed up, one in his struggles had laid hold on a plant of the Crithmum maritimum, which he providentially remembered *never grows beneath the surface of the water.* It was enough. Infinite Wisdom had 'cast the lot' for the Crithmum. It was the symbol of safety, and assurance was restored. Thus was the Rock Samphire, like the Olive branch to the Patriarch, fraught with the message of Providence."

On the south-east coast of England the Samphire-gatherer pursued a most dangerous work. Growing in crevices of the precipitous cliffs, only the practised eye and the daring foot could reach the plant. This was often done by letting down a man by means of a rope. Dr. Hamilton tells a touching incident of this plan of Samphire-gathering. Two men had lowered their companion from the top of the cliff by a rope tied round his waist. The desired plants had made their abode in a hollow of the rock, overhung by the precipice from which the rope was suspended. By means of a pole in his hand the man steered himself into the hollow, and gained a footing; but in the meantime the rope had grated against the edge of the cliff, and was worn through. Fearful was his situation, a fresh rope was lowered and swung to and fro, but it was some feet from him, held distant by the overhanging cliff; but how to reach it! Beneath him yawned a precipice of such dizzy height that he knew a false effort would be certain death. Yet to remain was equally certain death. Each vibration of the rope, as it swayed to and fro, brought it less near to him. The effort must be at once, or it would be hopeless. He made one desperate spring—and succeeded; the rope was in his hands, he clung for dear life, and his companions drew him to the summit. Dr. Hamilton compared that rope to Christ, and that urgent necessity to the need of the unpardoned.

The romance of Samphire-gathering is well known; we all feel a thrill when it is spoken of. Shakespeare has immortalised the calling :—

> " How fearful
> And dizzy 'tis to cast one's eyes so low!
> The crows and choughs, that wing the midway air,
> Show scarce so gross as beetles; half-way down
> Hangs one that gathers Samphire—dreadful trade!
> Methinks he seems no bigger than his head."

Fanny's specimens were more attainable—a kind-hearted naval captain who was of their party climbed up the rocks and procured a bundle of the plants, supposing she was going to make a pickle of them. The flowers open in August.

The Stone Parsley (Athamánta libanótis), is rare, it grows to the height of two feet, and bears hemispherical umbels, deeply pinnate leaves, and whitish flowers adorned with violet stamens. It frequents chalky hills.

Edward has specimens of both the Burnet Saxifrages from Kent. The common one (Pimpinélla saxífraga), much resembles the Earthnut, but is distinguished by having roundish leaflets on the root-leaves, while the segments of the stem-leaves are thread-like. The greater Burnet Saxifrage (P. magna), has all its leaflets oval and serrated. It grows freely in waste ground, while the common species frequents hilly pastures. Both flower in August.

We now come to the fifth group, characterised by having the smooth seed compressed at the sides. The first family in this group is that of the Water Parsnips. The Large-leaved species (Sium latifólium), with its handsome flat umbels, I have only seen in a print; but the Narrow-leaved (Sium angustifólium), is a common inhabitant of the ditches in the water-meadows in Wiltshire. The leaves are fairly large and pinnate; they are sometimes mistaken for those of the Water Cress, and being poisonous, evil effects accrue from the error. The terminal leaflet of the Water Cress is larger than any of the others, while the final leaflet of this Water Parsnip is smaller than its fellows. Attention to this distinction would save many a mistake. The umbels are small, and grow upon a stalk from the side of the stem opposite to the leaves.

The Procumbent Water Parsnip (S. nodiflórum), grows in the same ditches; it is easily distinguished by its prostrate growth and umbels seated on the main stem opposite the leaves.

The Creeping Water Parsnip (S. répens), I found in a similar locality, also in Wiltshire; it is a smaller plant, more slender in its habit—the bracts being large and numerous.

Marazion Marsh furnished Fanny with the Least Water Parsnip (S. inundátum). It is a very small slight plant, with pinnate leaves and minute umbels, containing about five flowers and placed on the stem in pairs rising out of the water. The leaves under water are thread-shaped.

The Whorled species with its Fennel-like leaves, inhabits salt marshes in the north and west. I have no specimen of it.

The Hedge Honewort and Corn Honewort (Sison amómum and ségetum), I have also sought in vain. The former has an umbel of about four rays or minute branches; in the latter the umbels droop, and the stem is Rush-like.

The Cowbane or Water Hemlock (Cicúta virósa), grows near Ely, and I have seen a pressed specimen from thence. The stem is furrowed, the leaves pinnate, each leaflet being cut in three lance-shaped divisions, and serrated at the edge. The umbel is large, and the flowers white. It contains a virulent poison, and is very dangerous for cows.

Here we have the true Hemlock (Conium maculátum). It is not a common plant. I gathered it near Little Ouseburn, a village in the neighbourhood of York. It is handsome, with bright green, repeatedly-pinnate leaves, a white umbel, and a beautifully shiny stalk spotted with purple. It contains a strong poison, which is, however, capable of being converted into a powerful medicine. The seed is the most poisonous part; dangerous mistakes have arisen from the similarity of this seed to that of the Anise, and articles explaining the points of distinction are found in medical journals. We look with interest at this plant as furnishing the poison of that greatest of non-Christian philosophers, Socrates.

The common Angelica (Angélica sylvéstris), grows abundantly in moist woods all over the kingdom; it is a stately plant, attaining a height of from five to eight feet. The sheathing leaf-stalks are remarkably large, and the umbels of flowers proportionately so. The full green of the twice-ternate leaves gives the plant a pleasing appearance; a tiny snail feeds upon the foliage, and takes its name from the plant. There is a garden Angelica which grows to a still larger size, the stalks of which make a delicious preserve.

The common Alexanders (Smýrnium olusátrum), have also ternate leaves on the stem; their root-leaves are twice-ternate. The umbels are of a middle size, and the foliage light green and smooth.

Fanny has brought the Wild Celery (Ápium graveólens, *Plate IX., fig.* 4), from the banks of the Looe river. The umbels are sessile in the axils of the branches. The plant is poisonous when wild, but becomes wholesome from cultivation.

The Wild Parsley grows on hedgebanks about Hawkhurst, in Kent; its bright green leaves and greenish umbels too closely resemble the garden Parsley to need a description. Its only classic or historic association is with the Olympic games.

> "Proud were the mighty conquerors
> Crowned in Olympic games,
> For they deemed that deathless honours
> Were entwined around their names.
> But sere was soon the Parsley leaf,
> And the Olive and the Bay;
> But the Christian's crown of Amaranth
> Shall never fade away."

The Fennel (Méum fœnículum, *Plate IX., fig.* 5), is the near relation of the Parsley, both as a condiment for the table and a plant honoured by classic superstitions:—

> "Above the lowly plants it towers,
> The Fennel with its yellow flowers;
> And, in an earlier age than ours,
> Was gifted with the wondrous powers
> Lost vision to restore.
>
> "It gave new strength and daring mood;
> And gladiators fierce and rude
> Mingled it in their daily food;
> And he who battled and subdued,
> A wreath of Fennel wore."

The plant has dark green leaves and yellow blossoms. Its

real association in the present day is with salmon; and it is a remarkable fact that on the coast of Cornwall, where the delicious fish called Salmon Peel abounds, the Fennel so desirable for its sauce and garnish is also abundant. I have also found it on waste ground about Warwick.

Tournefort tells us that the Ferula of the ancients is a species of Fennel; "it has a hollow stalk full of white marrow, which being well dried takes fire like a match. This fire holds a good while, and consumes the marrow very gently, without damaging the bark, which makes them use this plant in carrying fire from one place to another. Our sailors laid in a good store of it. This custom is of the greatest antiquity, and may help to explain a passage of Hesiod, who, speaking of the fire that Prometheus stole from Heaven, says he brought it in a Ferula. In all probability Prometheus made use of the pith of Ferula instead of a match, and taught men how to preserve fire in the stalks of this plant. The stem is strong enough to be leaned upon, but too light to hurt in striking; and, therefore, Bacchus, one of the greatest legislators of antiquity, wisely ordained the first men who drank wine to make use of this plant, because, when heated with drinking, they might break each other's heads with ordinary canes. The priests of Bacchus supported themselves on these stalks when they walked; and Pliny observes that the plant is greedily eaten by asses, though to other beasts of burden it is rank poison. We could not try the truth of this, there being only sheep and goats on the island. The plant is now used for making low stools; they take the dried stalks, and by placing them alternately in length and breadth, they form them into cubes, fastened at the corners with pegs of wood. These cubes are the visiting stools of the ladies of Amargos. Plutarch and Strabo take notice that Alexander kept Homer's works enclosed in a casket of Ferula, on account of its lightness; the body of the casket was made of this plant, and then covered with reck stuff or skin, set off with ribs of gold, and adorned with pearls and precious stones."

A specimen of the Spignel was given to me by a friend who found it in the Highlands. It has compact feathery leaves and whitish flowers (M. athamanticum).

The Goutweed (Ægopódium podagrária), I found in quantities about a farm called Spring End, in Yorkshire. It is a handsome plant, growing from two to three feet high; its flowers are greenish, and the numerous small umbels are arranged in a very large hemispherical general umbel. A decoction of this plant used to be much esteemed as an application for gouty limbs.

The Common Carraway (Cárum cárui), I have not found, but seen it in gardens. It has a branched stem, white umbels, and ribbed fruit. With the last we are familiar from the earliest infancy as the seed of our tiny sugar plums.

The Pepper Saxifrage (Cnídium siláus), grows also in Cornwall; it is an uninteresting plant, with flowers of a dull white, pinnate leaves, and an unpleasant smell.

The Hare's-ear family have bright yellow flowers. One of them, the Common Hare's-ear (Bupléurum rotundifólium), was sent to me from the coast of Wales last summer. It is a pleasing-looking plant, with brittle-branched stems and leaves, through which the stalk grows in the style which is called *perfoliate*.

The Slender-leaved species has a wiry stem and simple narrow leaves.

The blossoms of the Narrow-leaved Hare's-ear are reddish-cream colour; it is found in Devonshire.

The last species, the Falcate-leaved, is peculiar to Norton, in Essex, and is, doubtfully, wild.

The White-rot (Hydrocótyle vulgáris, *Plate IX.*, *fig.* 1), the last family of this group, is a little creeping plant, with shield-shaped leaves, and tiny umbels of pinkish flowers. Our specimens were gathered on the banks of the Loe Pool.

The Fool's Parsley (Æthúsa cynápium), was found by Edward in Kent. It is a poor, inexpressive, lean plant, with twice-pinnate small leaves, and a spare umbel of pinkish flowers. The most remarkable point is that the slender bracts turn straight downwards. It is poisonous, though so unattractive, and when eaten in mistake for true Parsley has very bad effects.

The Masterwort (Imperatória ostrúthium), I have never found; it has twice-ternate leaves, and alternate flower-stalks.

Nor has the Milk Parsley (Selínum palústre), come in my way. It is a tall plant, four feet high, with white flowers, and a rough furrowed stem.

The Lovage (Ligústicum), family is the last in this group. The two members appertain to Scotland and Cornwall. They have twice-ternate leaves and white flowers.

The Parsnip (Pastináca satíva), grows on the edges of fields among the chalk downs of Wiltshire. It has yellow flowers and glossy leaves. When cultivated, its root is very wholesome and nutritious; a curious thing, surely, that in this tribe some plants should be so good and some so bad, and that cultivation should remove the noxious qualities of some, and boiling those of others.

The Sulphur-wort (Peucedánum officinále), is the contribution of a Scotch friend, who procured it from the west coast. Its leaves are five times divided, the segments narrow, and the flowers buff.

The Hartworts (Tordýlium), are very rare, one of them doubtfully British; the flowers vary from white to pink, the umbels are far-spread, and the stem rough with bristles.

The Cow Parsnip (Herácleum spondýlium), so common as to be found in every meadow and hedgerow early in summer, has large flat umbels, the outer florets being much larger than those towards the centre. The petals are more or less pink; the large, dull green, pinnate leaves are much less divided than those of the generality in the Umbellifers.

The contradictory qualities of this order are thus accounted for in the "Penny Cyclopædia:"—All the poisons of umbelliferous plants are of an alkaline nature, and this quality is dissipated by heat: hence the change in their condition when boiled. The author of this article in the Cyclopædia points out what essentially good service this tribe does to us. The roots of the Parsnip and Carrot contain a large quantity of starch and sugar, which makes them very nutritious. The Celery, Sea Holly, and Angelica have pleasant stalks, which, when cultivated or cooked, are wholesome. The leaves of the Parsley, Fennel, and Chervil are also good when boiled. The thick juice of the Assafœtida, Hemlock, and others, though poisonous in itself, is useful as a medicine; and the seeds of the Coriander, Carraway, Dill, and Anise are very valuable. It occurs to me, that God permits poisonous plants to exist in order to compel us to use the senses He has given us, in avoiding what we find to be evil, and, by industry and research, remedying it, or turning it to beneficent purposes. All His works are good, and He teaches to the earnest and humble-minded how to find out the true and beautiful, and to gather wholesome fruit from the creation of His hand.

We are now so near the end of the Calyx order, that we had better look over the few remaining plants.

The Ivy family succeeds the Umbelliferous plants, and our native Ivy needs little description. Clothing the venerable ruin, and at once supporting it in its decrepitude, and adding to its beauty, the Ivy has continually been the theme of poets and moralists. It flowers in October, and its clusters of black berries are greatly enjoyed by the blackbirds and thrushes in spring. Sir Walter Scott describes it as mantling the rocks about Rokeby, where it does indeed thrive in great luxuriance at the present day.

> "Oft, too, the Ivy swathed their breast,
> And wreathed its garland round their crest,
> Or from the spires bade loosely flare
> Its tendrils in the upper air;
> And so the ivied banners' gleam
> Waved wildly o'er the brawling stream."

In old times it was worn around the brow at bacchanalian

revels, but in the purer age it was dedicated as a crown to the minstrel.

> "The Ivy meet for minstrel's hair."

The largest tree of Ivy (Hédera hélix), which I have ever seen grows at Fountains Abbey, the pride of our county. Its trunk measures three feet two inches round.

There is another family in this Ivy group, that of the Moschatel. Our only British member of this family is common in woods everywhere, flowering early in spring (Adóxa moschatellina). Its pale green flowers grow five in a head at the summit of the stalk, and a pair of three-parted leaves of a similar hue are situated half-way down. Other lobed leaves, each having three leaflets, spring from the root. The plant used to be called the "Gloryless," from its inconspicuous appearance, and its Latin name has a similar meaning.

The Wild Cornel group is the only remaining one in this order. One of our representatives is the Cornel or Dog-wood of our hedges. It is a bushy shrub, with egg-shaped pointed leaves, and clusters of cream-coloured flowers. The berries are small, and of a purplish-black. Skewers are made of the wood. The flowers are arranged in a small cyme. My specimen grew at the foot of Hungary Hill, near Ripon. I have also gathered it about Easby, and Edward has specimens from Hawkhurst. The bark of some foreign species of Cornel is a good tonic (Córnus sangúineus). The Cornel is mentioned by Homer as furnishing weapons of war:—

> "His Cornel spear
> Ulysses waved to rout the savage war."

I know not whether this be the same plant as ours, or even of the same family, but the specific name, *sanguineus* (bloody), seems to strengthen the idea that the tree furnished the spear-wood of the ancients. Indirectly it is still a warlike tree, for its wood makes the best charcoal for the manufacture of gunpowder. So well is this understood in Kent, that the landowners are jealous of planting the Cornel in their woods, because it is so constantly stolen and sold for the powder mills.

There is a Dwarf Cornel (C. suécica), which was sent to me from the Highlands; it is a mere herb, with large strongly-ribbed leaves, simple and broad, and a large four-petalled flower, cream-coloured, and with a deep stain of purple in the centre.

CHAPTER X.

FORMS OF COROLLAS—VISCÁCEÆ—CAPRIFOLIÁCEÆ—
RUBIÁCEÆ—VALERIANÁCEÆ—DIPSACÁCEÆ.

> "Flowers image forth the boundless love
> God bears His children all,
> Which ever droppeth from above
> Upon the great and small.
> Each blossom that adorns our path,
> So joyful and so fair,
> Is but a drop of love divine
> That fell and flourished there!"
>
> ILIMON.

WE have now gone through the Receptacle and Calyx subclasses, and are come to that of the COROLLIFLORALS, where the corolla is no longer formed of separate petals, but is all in one piece, or *monopetalous*. There are a variety of forms both amongst one-petalled and many-petalled corollas. The *Rose-shape* is one of the most familiar among the polypetalous flowers, with its five slightly-cupped petals (*Plate VII., fig.* 6). The *Cruciform* is equally well known, having only four petals arranged in the form of a cross (*Plate III.*). The *Lily-shape* is difficult to meet with among our wild flowers; the Wild Tulip and Wild Hyacinth are approaches to it (*Plate XVI.*). Of the *spurred* form we have examples in the Violet (*Plate IV., fig.* 3), Monkshood and Columbine (*Plate I., figs.* 9 and 10); and the *butterfly-shaped*, or *papilionaceous*, with its raised standard, spread wings, and curved keel, needs no description (*Plate VI.*).

The *one-petalled corolla*, which constitutes our present order, and has the stamens fastened to it, is susceptible also of several forms. The Speedwell (*Plate XIII., fig.* 6), Elder, Gueldres Rose (*Plate X., fig.* 2), and Yellow Bedstraw (*Plate X., fig.* 6), are examples of the *wheel-shaped*; the Campanula (*Plate XII.*,

L

fig. 1), and Linnæa of the *bell-shaped;* the Primrose (*Plate XIV., fig.* 3), and Anchusa (*Plate XII., fig.* 14), of the *salver-shaped;* the Honeysuckle (*Plate X., fig.* 3), of the *tubular;* the Bindweed (*Plate XII., fig.* 11), of the *funnel-shaped;* the Salvia (*Plate XIII., fig.* 8), Thyme, Mint, Dead Nettles, and Ground Ivy (*Plate XIII., fig.* 9), of the *labiate;* and the Snapdragon (*Plate XIII., fig.* 5), of the *gaping-shape.* This is the largest order of British plants, and many of our favourite flowers are contained in it.

The first order in the Corolla subclass is that of the Mistletoe (Víscum álbum, *Plate X., fig.* 1). Our one representative of this tribe is the well-known evergreen which holds so prominent a place in our Christmas festivities. It has its stamens and pistils on different flowers. Its seed is enveloped in a slimy covering, and is able, by its own adhesiveness, to remain fixed to the branches of trees. There is no calyx, and the corolla is four-cleft. The flowers and berries are both white. Plants of this tribe abound in tropical lands, adorning the trees, on which they live parasitically, with clusters of gay-coloured flowers.

Our interest in the Mistletoe is derived from ancient custom and habit. The Teutons had a great veneration for this plant. They held a curious tradition regarding it, which ran thus :— "One day Balder told his mother, Friga, that he should die. Friga conjured the elements, earth, air, fire and water, metals, maladies, trees, animals, and serpents, that they should do no evil to her son; and her conjurations were so powerful that nought could resist them. Balder, therefore, went to the combat of the gods, and fought in the midst of showers of arrows without fear. Loake, his enemy, wished to know the reason; he took the form of an old woman, and sought out Friga. He addressed her thus :—' In the midst of our fight the arrows and rocks fall on your son without hurting him!' 'I believe it,' replied Friga, 'all those substances are sworn to me. Nothing in nature can hurt him, except one little plant, growing on the bark of the Oak, with scarcely any root. I have not cared to ask its protection, for it is too weak to injure.' Loake immediately ran and found the plant, and gave an arrow formed of it to the blind Heda, telling him that Balder was before him. Heda discharged the arrow, and Balder fell, pierced and slain. Thus the offspring of a goddess was killed by an arrow of Mistletoe shot by a blind man." This legend is the origin of the respect felt both by the Gauls and Druids for this plant. I found the story in a book entitled "Forest Trees of Britain," by Mr. Johns.

The Mistletoe grows most frequently on Apple trees, where I have seen it in abundance in Somersetshire, and more sparingly in Kent and Devon. It is also found on the Pear, Hazel, Hawthorn, Maple, Willow, Ash, Elm, and Poplar. Dr. Borlase tells us that the Druids went in solemn procession to gather it from the Oak, to present it to Jupiter, with an invocation to all the world to assist at the ceremony.

" Mystic Mistletoe flaunted,
Such as the Druids cut down with golden hatchets at Yule-tide."

They had the Mistletoe in great veneration, and termed it the universal remedy. Two white bulls, never yoked, were led forth, and the priest, or arch-Druid, clothed in white, ascended the Oak, and cut the Mistletoe with a golden hook, while a white garment was spread beneath it, into which it fell. From this a kind of catholicon was formed, which cured all manner of diseases, and was an antidote to poison. The flowering time of the Mistletoe is April. Each seed contains the germ of two plants. When inserted into the bark of a suitable tree, it lies dormant for two years; after which the bark becomes fretted, and a young plant appears.

The Elder blossom (Sambúcus nigra), is a very familiar friend, the large flat cymes covering many a bush in every part of England about July, and the overpowering scent making itself perceptible at a great distance. High authorities are at variance about the excellencies of this bush. Evelyn says, that if its virtues were fully known no one need longer have any ailment; and Dr. Boërhaave, of Leyden, thought so highly of its qualities, that he never passed the tree without taking off his hat. But Pliny differed in opinion, and slightingly declared, in allusion to the hollow stems, that "the plant it was all skin and bones." Wine and syrup are made of the berries, both of which are very good for colds, and a great consolation to children for the annoyance of early going to bed. There is also a delectable wine made from the flowers. The "Elder ointment" is profoundly disbelieved in by the faculty, and declared to be even hurtful.

The Dwarf Elder (S. ébulus), has poisonous qualities; a decoction of its root, and one of its inner bark, has proved fatal. It is a smaller bush; I have heard of it near Edinburgh, and have a specimen from the neighbourhood of Darlington. The flowers are somewhat larger than in the common Elder, and they are tinted with lilac at the back of the petals.

The HONEYSUCKLE family comes next.

The pretty climbing Honeysuckle stands at the head of the family. Its fragrant tubular flowers, and simple, slightly-glaucous leaves, placed in pairs, are familiar to us, and scarcely less so are the small heads of brilliant crimson berries which succeed them. The Woodbine, or Honeysuckle (Lonicera periclýmenum, *Plate* X., *fig.* 3), is a favourite theme of poets.

> "But, from Flora's fairy realm,
>   Token would'st thou bring for me?
> Go where round yon towering Elm
>   Clings the Woodbine tenderly.
> Not to Fancy's ear alone,
>   Doth it kindly thought impart.
> Would'st thou soar, and strength hast none?
>   Clings to earth thy grovelling heart!
> Seek, like yonder fragile flower,
>   Fitting prop round which to twine;
> There's an Arm of love and power,
>   Lean on it, and heaven is thine."

I have a specimen of the Perfoliate Honeysuckle, but it is from a garden. The flowers of this species are always pale, and the stalk seems to grow through the middle of the leaf: hence it is called *Perfoliate*.

The Upright Honeysuckle, with its pairs of scentless flowers and distinct berries, is scarcely considered a native of Britain. Its leaves are simple and downy.

The common Gueldres Rose (Vibúrnum ópulus), is a frequent ornament of our groves. Its large cymes have small fertile flowers in the centre, and large barren ones round the edge. The leaves are lobed and cut. The whole shrub is smooth. The flowers appear in June; and in autumn the tree is very conspicuous, with its crimson leaves, and clusters of crimson berries. I remember our adjourning from a pic-nic to an extempore evening party, when I was in Wiltshire, and we transformed our plain muslin dresses into very elegant evening costumes by means of sprays of these berries that we gathered in a wood called Vallis. The Gueldres Rose is named from a district in the Low Countries, called Gueldres-land, where it flourishes in great abundance. Though to our taste the berries have a nauseous flavour, they are much esteemed by the Swedes, who make them into cakes with honey and flour. In Siberia a spirit is distilled from them.

The Mealy Gueldres Rose (V. lantána), I found on the Wiltshire downs. It has simple leaves, covered underneath with a thick yellow meal, and cymes of dull white flowers, all the same size. In some parts of the country it is called the "Cotton Tree," from its general mealiness. The berries are nearly black when ripe. Mrs. Howitt terms it the "Wild Hydrangea," and it certainly has some resemblance to that plant. In the Crimea the young branches are used for pipes, and they are employed in Germany in basket-making.

Now for Linnæus's favourite flower. This pretty creeper, with its pairs of rose-coloured bells, was discovered by him, frequenting Pine woods, where no other flowers could exist. My specimen was sent from the Dalmahoy Crags, near Edinburgh, where the plant is occasionally found. It is called after the great botanist, "Linnǽa boreális" (*fig.* 4). Miss Elliot has worthily celebrated its retreat:—

> " 'Tis a child of the old green woodlands;
>   Where the song of the free wild bird,
>   And swaying boughs of the summer breeze,
>   Are the only voices heard.
>
> " In the richest moss of the lonely dells,
>   Are its rosy petals found,
>   With the clear blue skies above it spread,
>   And the lordly trees around."

Our next British tribe is that of the MADDER, to which belong foreign plants of great importance. This contains the Ipecacuanha plant, a little shrub inhabiting the forests of Brazil, with a root about the size of a goose-quill, yellow, but with grey rings round it. This ringed bark forms a medicine equally appreciated by allopathic and homœopathic practitioners.

The Quinine, or rather the tree whose bark furnishes that valuable tonic medicine, is a member of the same order; and last, but not least, is their brother the Coffee plant. This is an evergreen shrub, a native of Arabia and Abyssinia.

Dr. Scoffern gives a legend of the first discovery of the use of the Coffee berry. The pious Mollah Chadelly was so afflicted that he could not keep awake during his nocturnal devotions, and besought Mahomet to reveal some means of keeping the sleep away. So Mahomet sent a herdsman to inform the devotee that his goats could not sleep after eating Coffee berries, but kept frisking about all night long. The Mollah, taking the hint, prepared a good dose of Coffee, and was charmed with the result. Not a wink of sleep did he get; delicious sensations crowded on his brain, and his midnight devotions were intensely fervent.

Another story is told relating much the same incidents, but making a Prior of a Maronite convent the hero. Afterwards the Turks became so devoted to Coffee, that they frequented the Coffee-shops instead of going to mosque, and so all use of the berry was forbidden on pain of stripes. But public opinion righted itself, and the Coffee became universally popular.

Our Madder seems insignificant compared with these plants of world-wide celebrity, but it, too, holds a highly respectable position. The order is sometimes called the Starry tribe, because the whorls of leaves are arranged in that form.

The Dyer's Madder is a foreign species of great utility; quantities of it were being used in some dyeing works which I visited near Manchester. The dye is extracted by a chemical process. It was cultivated in Normandy during the middle ages, and exported in large quantities. The French name for Madder, *Garance*, is derived from Verancia, the name applied by the Gauls, who, according to Strabo, cultivated it to some extent. The most curious circumstance relating to the Madder is, that if it is administered to any amimal in the state of decoction, it penetrates the whole organism, dyeing the very bones.

Our own Madder (Rúbia peregrína, *fig.* 5), furnishes a red colouring matter. Fanny found it on the banks of Loe Pool amongst brushwood. It has four or five leaves in each "star;" they are bright and evergreen, with bristles round them. The flowers are greenish-yellow, tubular above, and the segments of the corolla spreading into a wheel shape. It grows abundantly about Clevedon, as well as in Cornwall.

The Bedstraw family, which succeeds the Madder, has also the distinguishing mark of the Starry tribe; its whorls contain four, five, or six leaves. The flowers are wheel-shaped and four-cleft.

The Crosswort (Gálium cruciátum), is common everywhere. The flowers are arranged in small yellow clusters in the axils of the leaves, which are broad, and of a pale green. The strong scent of honey procures for it the title of "Honeywort" in some districts.

The Rough Marsh Bedstraw (G. uliginósum), has six leaves in a whorl, and very scattered clusters of clear white flowers at the end of its climbing stems. It grows in the ditches about Clevedon, clinging to the tall water plants.

The Yellow Bedstraw (G. vérum, *Plate X., fig.* 6), grows on the borders of fields and lanes, and about heaths everywhere. Its scent of honey is nearly as strong as that of the Crosswort, but its leaves are very small and narrow, and of a dark green; each whorl contains eight. The flower-cluster is very large, and much branched, forming the principal part of the plant, the colour deep yellow.

The great Hedge Bedstraw (G. mollúgo), is the handsomest species. One day when we were driving along the Great North-road, we were attracted by sheets of feathery whiteness on the hedge. We stopped, and I left the carriage, returning with branches of this beautiful plant. This was near Richmond, in Yorkshire, and I see that Edward has specimens from Kent.

The pretty Heath Bedstraw (G. saxátile), is a charming ornament of our woods and moors; many a rock is half covered with its thickly-set clusters of white blossoms. Its leaves are six in a whorl. The plant is prostrate, and much branched, but only three or four inches long. The seeds are grained, and tinged with crimson.

Edward has the Water Bedstraw, as well as the great Hedge Bedstraw, from the neighbourhood of Hawkhurst. It is a weak plant, with very small white flowers, and leaves four in a whorl.

The Goose-grass (G. aparine), has its leaves eight in a whorl, and its inconspicuous flowers are arranged in a small cluster, which springs from the axil of the leaf. Every part of it is beset with hooked bristles; it adheres by these to everything it touches, and is hence called by some "Cleavers." This habit makes it a favourite with mischievous children, who slyly attach portions of it to the dress of those they wish to torment. As little children, we used to push pins' heads into the pretty round bristly seed, and thus make "fancy pins."

The Northern Bedstraw resembles this, but its leaves are smooth, and only four in a whorl.

There are an Upright Bedstraw, with the leaves eight in a whorl, and dense white panicles; a Bearded Bedstraw, leaves six in a whorl, and seeds kidney-shaped; a Wall Bedstraw, leaves eight, and flowers few and pale; and a Least Bedstraw, with copious white flowers. Most of these are rare.

The Woodruff family comes next in the Starry tribe.

The Sweet Woodruff (Aspérula odoráta), is a charming little plant; its light, whorled foliage, and clear white flowers are uncommonly pretty. It is useful, too, for it not only yields a pleasant perfume when dried, but preserves woollen goods from moth. Withering was of opinion that an infusion of its dried leaves was far superior to Chinese Tea. It is a good substitute, too, for the Tonquin Bean in the preparation of snuff, not being, like it, prejudicial to the sight. Children spell its name in rhyme:

   Double U, double O, double D, E,
   R, O, double U, double F, E.

such being the old-fashioned style of orthography.

The Small Woodruff, or Squinancy-wort (A. cynánchica, *Plate X., fig.* 7), grows freely on the Wiltshire downs; and although it lacks the scent of the Sweet Woodruff, its pink clusters and dark green whorled leaves make it a great favourite with me. It continues in flower nearly all the summer, and is frequent in England, but never found in Wales or Scotland. It used to be esteemed as a remedy for quinsy: hence its popular name.

There is a blue Woodruff, but I have never seen it. Sir J. E. Smith thinks it is only an introduced plant, and gives Devonshire as the county of its settlement.

The Field Madder is the last family of the Starry tribe. Our one representative is a rough prostrate plant, much branched, and bearing a small umbel of lilac flowers in the bosom of the terminal whorl of leaves. I have gathered it often in corn fields in Wiltshire, and have seen it in similar localities in all other counties that I have visited. (Sherárdia arvensis, *fig.* 8).

The VALERIAN order comes next. There are foreign species which, when dried, are highly odoriferous. The Spikenard of Scripture is one of these. The leaves are variously used in medicinal preparations. Among the Hebrews it was a very costly drug: you remember that Judas valued the box of ointment of Spikenard, with which Mary anointed our Lord, at two hundred pence—about £6 9*s.* 2*d.* of our money. Spikenard was so much prized by the Romans, that Horace promises Virgil several dozens of wine for one box of Spikenard. They used to anoint their guests with it. In Turkey and Egypt two other species of Valerian, inhabitants of Austria, are used to perfume the baths.

The Red Valerian (Valeriána rúbra, *Plate X.*, *fig.* 9), grows plentifully on cliffs near the sea in many places. Our specimen came from Fowey, in Cornwall, where the gay rose-coloured bunches of flowers make a great show, enlivening the dark cliffs most charmingly. It has been sent to me also from Dawlish. The root of this plant is an excellent remedy for nervous diseases. De Théis thinks the name is derived from the Latin *valere*, to heal.

I have the Small Marsh and the Great Wild Valerian. The former attains the height of a foot only; its chief peculiarity is, that it bears its stamens on one plant and its stigmas on another. The corymbs of white blossoms, tinged with pink, are very pretty. The stem-leaves are pinnate, the root-leaves oval. It frequents the damp marshy meadows in Swaledale (V. dioica).

The great Wild Valerian (V. officinális), grows in moist bushy places. There is a valley near to Marske, in Yorkshire, full of it, and it is frequently to be met with in similar localities all over the country. It grows three feet high, with large corymbs of pure white flowers, and pinnate leaves, with lance-shaped segments.

These two species have three stamens. I omitted to mention that the flowers of the Red Valerian have only one stamen.

There is a Heart-leaved Valerian growing in Colinton Woods, near Edinburgh, which has rose-coloured flowers. I have not succeeded in procuring a specimen.

The second family in the Valerian tribe is that of the Corn Salad.

The Common Corn Salad, or Lamb's Lettuce (Fédia olitória), I found in May at the foot of Blackford Hill, near Edinburgh. It is a common-enough plant, with its pale blue clusters of inconspicuous flowers seated close to the leaves at the end of the stem. The leaves are tongue-shaped, and blunt. It is much branched, and the foliage is of a pale green. The plant was in former times cultivated for salad, and the French still call it "Monk's Salad"—*Salade de Chanoine.* Old Gerarde calls it "White pot herb," and says, "In winter, and the first months of the year, it serves for a salade herbe, and is with pleasure eaten with vinegar, salt, and oile, as other salades be, among which it is none of the worst."

Edward has the Toothed Corn Salad; he found it in corn fields about Hawkhurst, flowering in July (F. dentáta). Its flowers are lilac, and its leaves are toothed. There are a sharp-fruited and a carinated species of Corn Salad.

We now come to the TEASEL order. The Teasel plants have one petal, and a single-leafed calyx.

The most important member of the Teasel family is the Fuller's Teasel (Dípsacus fullónum, *Plate X.*, *fig.* 10). It grows to the height of six feet. The stems and leaves are prickly, and the head is beset with large hooks. This plant is very useful in cloth manufactories. It is dried in bundles, the heads turned outwards. I saw a number of these bundles prepared ready for the market, in a farm-house at Vallis, in Somersetshire, near which place they had been cultivated. The cloth is made to pass over the heads of the Teasel, which are sharp enough to raise the nap, but too elastic to tear the cloth.

Country people cut open the heads, where they frequently find a worm. An odd number of these worms, placed in a quill, is believed to charm away sickness. The Fuller's Teasel grows wild in many parts of Wiltshire. Its long serrated leaves grow in pairs, and their shoulders are joined together, so that they form a cup, where the rain and dew collect. This is the only British plant which has this habit of collecting water; but in tropical countries many trees are endowed with such a power. Dampier mentions that his thirst has often been allayed, when he was in the West Indies, by perforating the leaves of the Wild Pine, and receiving the water from below. Dr. Murray states, that in Africa there is a tree called Boa hollowed like a cistern, which receives and retains the dews. The Birchwort of Canada is supplied with two leaves resembling the halves of a bell, with a natural depression in the form of a spout; and these leaves afford a plentiful supply of water to animals, as well as the occasional traveller. Prince Maximilian, in his "Travels in the Brazils," informs us that the natives are well acquainted with dew-collecting leaves, and thus the wandering savages quench their thirst. The Pitcher-plants are described by Mr. St. John, in his explorations in Borneo, as such a beneficent contrivance for storing water. The Pitcher of the largest species, the Nepenthe Rajah, contains four pints. In one he found a drowned rat. The plants are as attractive for their beauty as for their utility. Dr. Murray thus concludes his remarks on the subject:—"These are indeed 'pools in the desert;' and it must forcibly strike the contemplative mind as very remarkable, that succulent plants should be found where we least expect to meet with them; that an Infinitely Wise Intelligence should have so formed vegetation as to minister its supplies, and so moulded its foliage as to become so many cisterns of water—living fountains—to quench the thirst of the great tide of animation which swarms in tropical countries, and without which they would certainly perish. In no region of the globe is there wanting traces of remarkable design: creation teems with such—it is full of miracles."

There is a Wild Teasel (D. sylvéstris), a less robust plant, with less rigid prickles, but we have not met with it.

The Smallest Teasel (D. pilósa), has pinnate leaves, or rather two little leaflets at the base of the principal leaf. The flower is white, and the head is roundish, not conical, like the larger species. It grows in damp places about Frome, on the borders of Somerset.

The Scabious family is the second in the Teasel order.

The Common Scabious has the corolla in four segments, the flowers round the edge of the flat head being large. It is of a lilac colour, and the leaves are pinnate and serrated. It grows abundantly on the borders of fields, and in chalk or limestone places.

The Smaller Scabious (Scabiósa columbária, *fig.* 11), is somewhat different; its corolla is in five segments. The root-leaves are ovate and notched, the stem-leaves narrow and pinnate. It grows on the downs near Warminster.

I have the Devil's-bit Scabious (S. succísa). Both the Blue and the White varieties grow in the hilly pastures in Swaledale. The root is abrupt, and an old legend declares that there was one portion of the root so healing in its powers that it could cure all maladies; and that the devil, enraged at mankind possessing so great a blessing, bit the healing part off. The head of the plant is round, and all the florets are of the same size.

CHAPTER XI.

COMPÓSITÆ.

> " Spake full well, in language quaint and olden,
> One who dwelleth by the castled Rhine,
> When he called the flowers, so blue and golden,
> Stars, that in earth's firmament do shine.
>
> " Wondrous truths, and manifold as wondrous,
> God hath written in the stars above ;
> But not less in the bright flow'rets under us
> Stands the revelation of His love ! "
>
> LONGFELLOW.

THE COMPOSITE flowers, which now come to us in the course of natural arrangement, form the largest of our orders. De Candolle, the celebrated botanist, divided these plants into groups :—1st, the *Strap-shaped* or Chicory group, where all the florets are strap-shaped and perfect ; 2nd, the *Tubular* group, where all the florets are tube-shaped and perfect ; 3rd, the Corymb group, where the florets of the centre or *disk* are perfect, and those of the margin or *circumference* are strap-shaped, and have pistils only. All the flowers of this tribe are composed of numerous florets, as indeed the name of the order, " Composite," indicates. Lindley computes the number of genera in this tribe at 1005. They are mostly herbaceous plants, but some few attain to the stature of trees. The king of these is the Synchodendron, a tree fifty feet high, a native of Madagascar. The appearance of its flowers is a signal to the inhabitants to sow their Rice.

The Strap-shaped group has some useful members. The Chicory leaves are good as a salad, and the powdered roots are mixed with Coffee. Endive is another kind of Chicory, but not a wild species. The root of the Dandelion makes a wholesome tonic medicine.

The Tubular group boasts the Artichoke and the Tarragon for articles of food, and the Wormwood and Matico for medicines.

The Corymb group has only the Jerusalem Artichoke for eating; but it is rich in medicines, possessing Camomile as a tonic; Coltsfoot for coughs, and Calendula, one of the most useful homœopathic medicines for external application; and Feverfew as a bitter. A yellow dye is procured from the Corn Marigold, and carmine from the flowers of the Dahlia.

The Goat's-beard is the first family in the Strap group. The involucre has one row of long scales; all the florets are strap-shaped, and the seeds are long and feathery, forming a large handsome globe. We have three British species.

The Yellow Goat's-beard (Tragopógon praténsis), grows pretty frequently in meadows. Its stem is tall and few-flowered, and the leaves are long and tapering. I have gathered it both in Yorkshire and Durham. All the Goat's-beards close early; the strap-shaped florets rise into a perpendicular position, and the segments of the involucre draw together. This species used to be called "Jack-go-to-bed-at-noon." Linnæus called this closing "the sleep of the plants;" it is very fully shown among the Compound flowers. The Hawkweed Picris shuts at noon, the Chicory and Nipplewort in the evening, the Cat's-ear at four o'clock, and the Hawkweeds about three o'clock. Marvell speaks of this floral dial:

> "How well the skilful gardener drew,
> Of flowers and herbs this dial new!
> Where from above the milder sun
> Does through a fragrant zodiac run,
> And, as it works, th' industrious bee
> Computes its time as well as we.
> How could such sweet and wholesome hours
> Be reckoned but with herbs and flowers?"

The Purple Goat's-beard closely resembles the yellow species, except in the colour of its petals, and the greater length of the segments of the involucre. My specimen is from a garden. It used to be much cultivated for its edible roots, and is still to be found in gardens of a high order. The herb is called Salsafy.

There is a rare species of this family called the Greater Goat's-beard (T. májor), and distinguished by the swollen flower-stalks. It appeared upon the Calton Hill, Edinburgh, a few years ago; fresh earth was thrown over the turf afterwards, and the Goat's-beard plants have never since attained their former height of a foot and a half, but are stinted to little more than half the size.

Kent furnishes the Ox-tongue; a sandy bank on the Junction-road, near Hawkhurst, produces dozens of plants. It grows two or three feet high, has bristles everywhere, and there is an outer calyx of five heart-shaped lobes. The flowers are small and yellow, and the leaves lance-shaped. It is called Ox-tongue on account of its roughness (Picris echioides).

The Hawkweed Picris is a common autumn flower; the stem-leaves waved, the root-leaves toothed; the whole plant very rough, but not bristly, and the blossoms larger than those of the Ox-tongue.

The Hawkbits are in my collection; two of the species at least.

The Rough Hawkbit (Apárgia híspida), bears only one flower on a stem, and no leaves. The flowers are hairy at their opening, and the leaves, which spring from the root, are notched, and the points of the scallops turned back. I gathered my specimen on the top of Scarth Nick, a commanding hill in Wensleydale.

The Autumnal Hawkbit (A. autumnális), is an elegant little plant, with a branched stalk, bearing several bright yellow flowers. The partial stalks are scaly; the leaves smooth, and very much notched.

The Deficient Hawkbit, or Hairy Thríncia, has its florets red beneath, and is thus distinguished from the Rough species (Thríncia hírta).

The Dandelion Hawkbit is also single-flowered, but its leaves are smooth.

The Cat's-ear is the next family. The Long-rooted species is common in our pastures, flowering early in the autumn. The leaves are *runcinate*—that is, the lobes are pointed and turned back, and the stems are branched and smooth. The flower is large, and bright yellow (Hypocháris radicáta).

The Spotted Cat's-ear has its stem solitary, and its leaves rough, and blotched with purple; and the smooth Cat's-ear is without hairs, and the stem branched and leafy.

The Hawkbits and Cat's-ear are all small plants, growing less than a foot high.

The Sow Thistle family, the next in succession, are of a much higher growth.

The common Sow Thistle (Sónchus oleráceus), a troublesome garden weed, is stout, succulent, and milky. The heads grow in clusters, and the involucres are smooth. The leaves are runcinate, with round ears close to the stem. We used to seek it diligently as rabbit food.

The Corn Sow Thistle (S. arvénsis), is a stately plant, growing six feet high, and lifting its clusters of large golden flowers above the ripe corn. The stem-leaves are oblong and notched, the lower ones pinnatifid. This is as showy a plant as the class contains, if we except its brother, the rare Marsh Sow Thistle, which grows to a still greater height, having the leaves arrow-shaped at the base.

In mountainous districts a Blue Sow Thistle is found, with a brown involucre. I have no specimen of it.

One of the Lettuce family is in my hands, the common Wall Lettuce (Lactúca murális). It is a tall slender plant, growing in groves and hedges. The leaves are strongly lobed and runcinate, and often much tinged with purple. It flowers in a scattered panicle, each stalk being placed at right angles from that from which it springs. The flowers are formed of five pale yellow florets.

The Strong-scented Lettuce (L. virósa), grows in basaltic rock near Edinburgh. It has a finely-toothed oval leaf, a fetid smell, and clusters of pale yellow flowers. The plant is bitter to the taste, and has poisonous qualities.

The Prickly Lettuce (L. scarióla), carries its leaves perpendicular, whilst the Strong-scented Lettuce holds its leaves horizontally.

The Least Lettuce (L. saligna), has pinnate leaves.

The Garden Lettuce is quite a different species; it is not here, as in the case of the Wild Celery, that cultivation makes all the difference between the poisonous and the wholesome plant.

The Smooth Hawk's-beard (Crépis vírens), is my only species of that family. Its root-leaves are runcinate; the stem-leaves lanceolate and toothed, without footstalks. It grows on waste places, and has very small yellow flowers. The stem is much branched.

There are a Rough Hawk's-beard, with a downy involucre; and a Stinking Hawk's-beard, with hairy pinnate leaves, smelling like bitter almonds. These two inhabit chalky ground.

The Small-flowered Hawk's-beard (C. púlchra), is an elegant plant, having a corymb of small yellow flowers, and root-leaves oblong, while the stem-leaves are arrow-shaped.

The Hawkweed family is the largest in the Strap-shaped group. There was a wide-spread superstition in old times that the hawk fed its young ones on these plants, and that the birds owe their keenness of vision to them. In France, Germany, and Italy they are called Hawk's-plants.

We have two species, with naked stems springing from the root, and bearing a single flower. One of these is the pretty Mouse-ear Hawkweed (Hierácium pilosélla), that covers large plots in hedgebanks with its creeping stems, pale, oval, woolly leaves, and lemon flowers lined with crimson. It is common all over Britain. My specimens are from Yorkshire.

The Alpine Hawkweed (H. alpínum), varies in having a golden bloom.

There are three British species, with branching leafless stalks, but all belong to mountains.

The Branching Mouse-ear Hawkweed (H. dúbium), has lemon flowers, like its namesake, but its leaves are more pointed, and the branching stalk is an easy mark of distinction.

The Orange Mouse-ear and Orange Hawkweed (H. aurícula and aurantiacum), resemble each other; but the latter is larger, and has entire leaves, while those of the former are slightly notched.

Of the Hawkweeds with leafy stems we have a great number.

The Wall Hawkweed (H. murínum), grows about Easby; it has one leaf close to its corymb. Its leaves are toothed, and its flowers large and yellow.

The Stained-leaved Hawkweed (H. maculátum), has dark green leaves, spotted with black.

The Wood Hawkweed (H. sylváticum, *Plate XI., fig.* 1), carries its flowers in upright racemes. Its leaves are ovate and pointed, and the florets pale.

The Shrubby Hawkweed is very leafy and erect. The leaves clasp the stem, are ovate and pointed, and sharply toothed. This is a common species, growing in woods and hedgerows in the autumn.

There are a Lungwort Hawkweed, with leaves clouded with purple; a Glaucous Hawkweed, with large lemon flowers; a Marsh Hawkweed, with an angular stem; a Soft-leaved Hawkweed, of slender growth and golden flowers; a Shaggy Hawkweed, whose lemon flowers are nearly solitary; and a Narrow-leaved Hawkweed, carrying its flowers in umbels. All these are to be found among the Scotch mountains. The time of closing of the Hawkweed flowers forms a part of the floral dial.

> "See Hieracium's various tribes
> Of plumy seed and radiate flowers,
> The blooms of time their course describe,
> And wake and sleep appointed hours."

We now come to a very familiar plant, the Dandelion (Leóntodon taráxacum), so called from "Dent de lion," because the cutting of the leaves resembles lions' teeth. Elliot calls this flower the "Sunflower of the spring," and the French term it, for its seed's sake, "Couronne de prêtre." The poet Lowell thus celebrates its praise:—

> " Dear common flower, that grow'st beside the way,
> Fringing the dusty road with harmless gold,
> First pledge of blithesome May,
> Which children pluck, and, full of pride, behold,
> High-hearted buccaneers, o'erjoyed that they
> An El Dorado in the grass have found,
> Which not the rich earth's ample round
> May match in wealth! Thou art more dear to me
> Than all the prouder summer blooms may be!"

We used to make chains of the stalks when we were children, but we always stained our pinafores with them, and got into trouble. The best use of them, in our opinion, was, that we had a penny a-hundred for gathering their blossoms off the lawn. The leaves when blanched are used as a winter salad on the continent, and a valuable medicine is obtained from the root.

The Nipplewort family comes after the Dandelion. There are two British species.

The common Nipplewort (Lapsána commúnis), bears its small yellow flowers in panicles, somewhat resembling the Wall Lettuce, but branching at acute and not right angles. The leaves are ovate, stalked, and toothed.

The Dwarf Nipplewort (L. pusilla), has leafless stalks and pale flowers. The root-leaves are rough-edged.

The common Nipplewort grows abundantly in our woods, hedges, and waste places, but we have not the dwarf species.

A pleasing contrast among this crowd of yellow "stars" is the Blue Chicory (Cichórium íntybus, *Plate XI., fig.* 2), with its woody stem and plentiful flowers. It is a very attractive plant. Fanny found it growing by the roadside near Clevedon, and also about Looe, in Cornwall. The Germans call it "Keeper of the Ways." Its flowers open at eight o'clock in the morning, and close early in the afternoon. It is also mentioned in the lines from which I have already quoted:—

> "On upland slopes the shepherds mark
> The hour, when to the dial true,
> Cichorium, to the soaring lark,
> Lifts her soft eyes serenely blue."

Its utility as an adjunct to Coffee has already been named; it has a classical association in being one of the articles of diet used by Horace. It is cultivated as a substitute for its ally, Endive, in English gardens, and to a much greater extent in Flanders and Holland, where salad forms a more important article of diet. This is the last family in the Strap-shaped or Chicory group.

We come next to the Tubular group, but this needs to be subdivided. We will take first those tubular plants which are arranged in a round head, like the Thistles; and afterwards those that are in a flat head, like the Tansy. In the Strap-shaped group the most of the flowers were yellow; in this first division of the Tubular group we shall see lilac the prevailing colour.

The Bur Dock is the first family. We have only one species. The common Bur Dock (Arctium láppa), is a large coarse plant, often growing by the roadside, and covered with dust. Its dingy lilac flowers are very uninteresting, and the hooked bristles on its involucre make it very troublesome in adhering to the dress. The root of this plant is said to be good for rheumatism.

The Saw-wort (Serrátula tinctória), is abundant both in Kent and Somerset. It is a slender plant, with dark green, smooth, pinnate leaves, and branched clusters of purple flowers. I see that Fanny has specimens from Clevedon. It flowers in July and August.

The Alpine Saw-wort (S. alpína), has pink flowers and simple leaves. I hope to find it some day.

Now for the Thistles. The Musk Thistle (Cárduus nútans),

is certainly as pretty as any. Its flowers are of a crimson purple, very large, solitary, and drooping; the scent is agreeable. I gathered my specimen on a wild moor in Swaledale, near the Old Gang Lead Mine.

The Welted Thistle (C. acanthoídes), is very prickly. Its heads are clustered generally in threes, and the stem is winged and thorny. Its flower is smaller than that of the last species. It is common in waste places.

Here is the Slender-flowered Thistle (C. tenuiflórus). It likes waste ground near the sea. Fanny found this specimen on the East Hoe, Plymouth. There is cotton about the stem and involucre, and the crimson flowers are very shabby. Some are very pale; indeed all the Thistles are apt to vary to white.

The handsomest species, in my opinion, is the Milk Thistle (C. Marianus), with its tall stems, large flowers, and broad bright leaves, veined with white. The legend is, that as the Holy Family travelled into Egypt, some milk fell upon this plant, and produced the white veins. I found my specimen at Monkton Deverill, in Wiltshire.

The Lance Thistle (C. lanceolátum), is also handsome. Its head is comparatively solitary. It grows in Kent and elsewhere. It is a tall erect plant, and very prickly.

The Woolly-headed Thistle (Cnícus crióphorus), is both curious and beautiful. It grows in waste places. I gathered my specimen from a railway embankment, near Leamington. Its leaves are bristly above, and cottony below, and the numerous bracts constituting the involucre are interwoven with wool, like cobweb.

The Creeping Thistle (C. arvénsis), is a common weed, growing in waste places. It is less offensive than its fellows, having no prickles about its clusters of pale lilac flowers.

The Marsh Thistle (C. palústris), grows freely in damp places in every district. It has crimson flowers in small clusters, closely seated on the stalk.

The Dwarf Thistle (C. acáulis), I got on the Wiltshire downs.

It has pretty dark prickly leaves, veined with pale yellow or white, and handsome flowers growing quite close to the ground.

A Tuberous Thistle (C. tuberósa), grows in Greatridge Woods, in the same neighbourhood, so called from the form of its root. It has recently been found, too, near Cirencester, but I have failed in procuring a specimen.

A dried specimen of the Meadow Thistle (C. praténsis), has been given to me from Essex. It is a smaller plant than any of the Thistles, except the Dwarf. The flower is solitary, the stem naked and woolly, and the leaves lance-shaped and wavy.

The last of this family, the Melancholy Thistle (C. heterophýllus, *Plate X., fig.* 3), grows in great abundance in meadows in the hilly country of Yorkshire. We were much puzzled on first observing the clusters of large succulent leaves, lined with white, among the grass in the meadows and pastures. Presently a short stem appeared, with a round bud; the stalk lengthened, until it reached the height of four feet, one or two wavy leaves adorning it; then the multitude of purple florets composing the splendid head opened. The stem is slightly bowed, occasionally once-branched, and it has a soft bloom upon it. Like many of its brethren, it is very fragrant as well as very beautiful. It flowers in July.

A first cousin of these true Thistles is the Scotch Thistle, or, as it is often called, the "Cotton Thistle" (Onopórdum acánthium). This is the plant whose vigorous spines pierced the foot of the invading Dane, and wrung from him the cry which betrayed his intended midnight assault. It is a sturdy plant, studded everywhere with sharp thorns, and the stem and underpart of the leaves are covered with cotton. I found my specimen close to a quarry of lias stone, near Leamington. As I meditated securing a portion of the plant, its whole appearance seemed to utter the saying, "Wha daur meddle wi' me?"

The Carline Thistle (Carlína vulgáris), is the next family. It is a dingy yellow colour, low of growth, and very prickly. It abounds on limestone pastures. The flowers do not decay; but after the plant itself is shrivelled and colourless, the yellowish flowers still dot the turf. My specimen is from Applegarth, near Richmond.

The Black Knapweed (Centaúrea nígra), the most common of the Knapweed family, is familiarly known as Hardhead. The flower is like a small Thistle, but the plant is smooth and tough. In this, the next species, the scales of the involucre are fringed with dark bristles. This is the family whose Latin name is derived from the Centaur Chiron, who cured his foot of the wound inflicted by Hercules, by applying one of these plants. They used to be thought highly medicinal, but I fancy no one believes in them now.

The Greater Knapweed (C. scabiósa), is a handsome plant. The florets round the edge are very large, and the colour is a bright purplish-crimson. The leaves are pinnate.

The Corn Bluebottle (C. cyânus, *Plate XI.*, *fig.* 4), is the prettiest. Fanny gathered it near Penzance. Its brilliant colour entitles it to its Scotch appellation of "Blue Bonnet." The French call it "Bluet." Surely this must be the plant meant by Longfellow, when he says,

> "Others, their blue eyes with tears o'erflowing,
> Stand, like Ruth, among the golden corn."

The Star Thistle (C. calcítrapa), is as prickly in its involucre as that of any plant of the family after which it is named. Its leaves are spreading, and the rose-coloured flowers spring from the side of the stem.

The Jersey Star Thistle (C. Isnárdi), another of the Knapweeds, has a spiny involucre, but its flowers are terminal. In travelling from London to Gravesend I once saw the yellow Star Thistle, with its lyre-shaped root-leaves and terminal solitary flowers. I would have given much to have stopped for a specimen.

The Brown Knapweed (C. jácea), resembles the Hardhead, but its florets are more spread. It is very rare. This family concludes the first division of the Tubular group.

The second division has *flat heads*.

The Bur Marigold is the first family in this division. We have two British species.

The Nodding Bur Marigold (Bídens cérnua), I found on damp ground at Hill Deverill, in Wiltshire. The involucre is in large segments, which extend far beyond the dark yellow head. The leaves are lance-shaped and serrated, and the flowers drooping. The plant grows two feet high.

The Three-leaved Bur Marigold (B. tripartíta), much resembles this; its principal difference is expressed in its name. It grows near Reading.

Most watery places furnish the Hemp Agrimony (Eupatórium cannabínum). Edward describes it as flourishing luxuriantly in Kent, and I have found it in abundance on the banks of the Ure, near Ripon, and on the margin of streams about Richmond and Bedale. Its leaves are in many segments, like the fingers of a hand. It grows four feet high. But though its corymbs are large enough, and each little head neat enough, it is quite spoiled by its dirty lilac colour. If the tint of the flowers were as pretty as that of the leaves, it would be a very handsome plant.

I have a specimen of the Goldilocks (Chrysócoma linosýris), the next family to that of the Hemp Agrimony, but it is not a wild one. It was given to me by Mr. Ward, and grew in his garden. The form of the plant is wand-like. The main stem and numerous flower-stalks are beset with narrow glaucous leaves. The heads are yellow.

The Cotton Weed (Diótis marítima), is an inhabitant of sea-shores. The leaves are oblong and notched, and the yellow flowers are placed in corymbs. The whole plant is covered with cotton.

The Tansy (Tanacétum vulgáre), is very familiar to me, as growing along with the Hemp Agrimony by the Ure. Its heads are flat, yellow, and in large corymbs, and its leaves are cut into tiny leaflets. The whole plant has a strong aromatic smell and flavour, and in old times was made into a pudding on Easter day, the custom, doubtless, having its origin in the "bitter herbs" of the Passover.

After the Tansy family comes that of the Wormwood.

The common Mugwort (Artemísia vulgáris), is a member of its family. It is a tall plant, with twice-pinnate leaves, dark green above, and white below. The heads are ovate, and of a brownish colour. It grows in hedges and waste places. The common people used to have a superstitious reverence for this plant, and believed that so long as they carried a piece of it about with them they would feel no fatigue. It is used instead of Hops for making beer in Sweden. My plant grew near little Ouseburn, in Yorkshire.

The common Wormwood (A. absinthium), grows about villages in Yorkshire, and abounds in cottage gardens. The leaves are all clothed with silky down, and the round heads of flowers droop. The whole plant has a bitter scent. It is still used as an emetic medicine in country places. Last year I was very ill for some time, my worst symptoms being violent and continuous sickness. The day after this was allayed, a stout servant girl arrived from a neighbouring farm, with "Missus's respects, and if the lady will take a pint of Wormwood tea, it will be sure to cure her outright." A large bundle of the herb was presented along with the message.

The Field Wormwood (A. campéstris), is peculiar to the eastern counties. The stem is procumbent, and it is scentless.

The Upright Sea Wormwood (A. gállica), was given to me by the obliging curator of the Edinburgh Botanic Gardens. It closely resembles the common Wormwood.

Fanny found the Sea Wormwood (A. marítima), or Southernwood in a salt marsh near Clevedon. The foliage is covered with white silky hairs, and the plant has a very aromatic scent. The flower-clusters droop, and the heads are pale and inconspicuous. This is closely allied to the "Lads' love and Lasses' delight" of village gardens.

We now come to the Cudweed family, and the first member of it is from Edward's Kentish collection.

The Bog Cudweed (Gnaphálium uliginósum), grows in damp clayey ground. Its flowers are brownish, and the stem much branched. The leaves are narrow, and the entire stature of the plant seldom exceeds four or five inches.

Another of this family has similar foliage. It grows in stubble fields, on a clay soil. About two inches from the ground a round head of flowers appears, of a hue resembling the leaves. From the sides of this head two or three branches shoot, bearing more leaves and heads of flowers. This is the common Cudweed, called by old herbalists "Herba impia," because it gave the idea of children trying to domineer over their parents (G. germánicum).

Yorkshire contributes two species to the Cudweed family, both inhabitants of my favourite valley of the Swale.

The Mountain Cudweed (G. dioícum, *Plate XI., fig.* 5), is a pretty plant, with small heads of flowers, in pinkish involucres, arranged in corymbs. The male flowers grow on one plant, and the female on another. The whole height does not exceed four inches. The leaves are ovate, green on the upper, white on the under surface, and the habit of the plant is creeping.

The Highland Cudweed (G. sylváticum), inhabits the more wooded parts of the same hills, over the exposed surfaces of which the Mountain Cudweed spreads its corymbs. The stem of this species is simple, and the flowers are arranged in a compound leafy spike. The leaves are of a narrow lance-shape.

The Pearly Everlasting (G. margaritáceum), is cottony all over, and has a large corymb of white flowers. Though it is sometimes found wild, yet it is better known as a garden plant. My specimen was found, apparently wild, in Swaledale.

The Least Cudweed (G. mínimum), closely resembles the Marsh Cudweed, differing chiefly in size, and in having broader leaves. It is a common plant in clay ground.

The Jersey Cudweed (G. luteo-álbum), has a yellow involucre, and the Alpine Cudweed has very slender stems.

The Butter Bur family concludes the second division of the Tubular group.

Our one British species (Petasítes vulgáris), with its thick spikes of dingy lilac blossoms and large kidney-shaped leaves, is too familiar to need description. It is the torment of the farmer in wet lands, spreading over acres of ground, and destroying all other vegetation. It used to be considered a sovereign remedy for the plague, and called, in consequence, "Plague-Flower." In memory of this custom Anne Pratt writes:

> "No gem-like eye glitters in thy pale face,
> No rich aroma breathes from thy dull lip;
> Yet, Petasites, there is that in thee
> Which calls emotion from its lurking-place.
> There is a scene to which thou art allied—
> A room the sun scarce sees; an atmosphere
> Converted into poison, on the couch
> The plague-spot marks his own."

In the group upon which we have now to enter, the Corymb group, the florets are generally of two colours; those of the centre yellow, and the rays white. There are many exceptions with yellow rays, and some with purple.

The Coltsfoot is the first family. Our one British member (Tussilágo fárfara, *fig.* 6), puts up its starry, sweet-scented flowers early in the spring. Each stalk is crowned by one flower, with the centre and rays bright yellow. The leaves do not appear for some time after. The stems are adorned with pinkish scales. The Highlanders use the leaves, when dried, as tinder, and a decoction of the flowers is an excellent remedy for coughs. The plant is very common on moist clay lands.

The Fleabanes are pretty, slender plants. The Canadian Fleabane (Erígeron canadénsis), grows abundantly upon every old wall about Bath, and on rocks and walls in that neighbourhood.

The Blue Fleabane I have never found.

The Michaelmas Daisy (Áster tripólium), adorns the salt

marshes about Clevedon with its fleshy leaves and clusters of purple flowers, and flourishes in the damp rocky places about the Lizard Cliffs. It is sometimes called "Sea Starwort," and is eagerly welcomed, because it blooms at a season when other flowers are scarcer. This is our only British Aster.

Our gardens present beautiful foreign species, richly-coloured and double, most of them from China or Germany. A tall species, frequent in cottage gardens, and popularly known as the "Michaelmas Daisy," I gathered wild on the banks of the Neckar, in Germany; and a more beautiful one, of lower growth and large blue blossoms, I brought from the Jura Mountains.

The Golden Rod (Solidágo virgáurea), is abundant in Kent: it grows freely in the hedges about Hawkhurst. We find it pretty often in Yorkshire, but it is certainly a less common plant there. It is tall and wand-like, and has perpendicular branches, and flower-stalks springing from them. Both the flowers of the disk and the rays are yellow. It used to be valued as a cure for wounds, and was sold in the London markets by the herb-women. It flowers from August to October.

The Ragwort family succeeds that of the Golden Rod. It has numerous members scattered all over the globe; in all they number six hundred.

The common Groundsel (Senécio vulgáris), so well known as a bird food, needs no description. When water is poured upon it, and allowed to stand awhile, it forms a softening wash for heated skin. This species has no marginal flowers or *rays*.

The Mountain Ragwort (S. sylvática), which I found on the sandy hedgebanks in Cheshire, and Fanny on the cliffs at Polperro, has small rays, but they curl back soon after opening. Its habit resembles that of the Groundsel, but the flowers do not droop, and it is much taller.

The Inelegant Ragwort (S. squálidus, *Plate XI.*, *fig.* 7), is a graduate of Oxford, growing on old walls about that notable place. The appearance of this plant belies its name; for though its branched stem is somewhat straggling, its golden flowers are handsome and well rayed, and its form has some grace about it.

The Hoary Ragwort (S. tenuiflórus), is abundant about Hawkhurst. Tall, downy, with divided leaves, and large clusters of gay yellow flowers.

The common Ragwort, which our country people call "Haygreen," or "Dogstander" (S. jacobǽa), is a common weed. Its toughness makes it difficult to remove. It greatly resembles the Hoary Ragwort, but is shorter, and its leaves of a brighter green.

The Marsh Ragwort (S. aquáticus), has purple stems and fewer flowers. It is a more elegant plant. Both these species are exceedingly common.

There is a Stinking Ragwort (S. viscósus), which grows in great abundance on the slopes of Arthur's Seat, and many other places about Edinburgh. It is very like the Groundsel, and grows as wildly, where it once settles itself, as that plant does in a forsaken garden. It is extremely sticky, and has a faint smell.

The Green-scaled Ragwort is so like the Mountain Ragwort, that it can only be distinguished by the green tips of its bracts, those of the Mountain species being coloured with blackish-purple.

The great Fen Ragwort (S. paludósus), was sent to me from Westmoreland last year. It is a noble plant, from three to six feet high, with large clusters of golden blossoms. Its leaves, and those of the next species, are undivided.

The Broad-leaved Ragwort (S. saracénicus), is fully as tall a plant, but its leaves are broader, and its rays more conspicuous. My specimen grew in the Botanic Gardens at Edinburgh. It is disappointing, in pressing these plants, to find that the seed will go on to ripen, and lose all the beauty of the yellow corolla.

The Field Fleawort (Cineráría integrifólia), is abundant on the Wiltshire downs. A pretty little plant, with simple woolly leaves and a naked stem, about four inches high, bearing two or three bright yellow flowers at the top.

The Marsh Fleawort (C. palústris), is a much larger plant, with toothed leaves. It is peculiar to the eastern counties.

The Great Leopard's-bane (Dorónicum pardaliánches), is a handsome plant, three feet high, with large yellow flowers, but Sir J. E. Smith says it is not indigenous.

The Ploughman's Spikenard (Conýza squarrósa), I found in October, at Vallis, on the borders of Somersetshire. It is a woody plant, with notched lance-shaped leaves, and the scales of the involucre turned back. The flower is a brownish-yellow. It is the only British member of the family.

The family of the Elecampane comes next. The common species (Ínula dysentérica), I found first in a moist field at Aldfield, near Ripon; I have since seen it pretty frequently. Edward says it is very abundant on roadsides and pastures in Kent. The leaves are oblong, very woolly, and clasp the stem, which is branched. The flowers are large and golden. It has a sweet musky smell.

The Great Elecampane (I. helénium), is very rare. Its leaves are rugged, but only downy beneath, and its flowers are very large and handsome.

The Small Elecampane has pale yellow hemispherical flowers, while the Samphire-leaved species is distinguished by its narrow fleshy leaves, and the orange hue of the florets of the disk. It grows on Portland Island.

At last comes the turn of our common Daisy (Béllis perénnis, *Plate XI., fig.* 8). No description of this is necessary. The French call it "Marguerite," because they liken it to a pearl; and Chaucer terms it "the ee of daie." Wordsworth celebrates its praise:—

> "When soothed awhile by milder airs,
> Thee Winter in the garland wears
> That thinly shades his few grey hairs;
> Spring cannot shun thee:
> Whole Summer-fields are thine by right;
> And Autumn, melancholy wight!
> Doth in thy crimson head delight,
> When rains are on thee.

> "In shoals and bands, a morrice train,
>   Thou greet'st the traveller in the lane;
>   If welcomed once thou count'st it gain:
>   Thou art not daunted,
>   Nor car'st if thou be set at nought;
>   And oft alone in nooks remote
>   We meet thee, like a pleasant thought,
>   When such are wanted.
>
> "Be Violets, in their secret mews,
>   The flowers the wanton zephyrs choose;
>   Proud be the Rose, with rains and dews
>   Her head unpearling;
>   Thou liv'st with less ambitious aim,
>   Yet hast not gone without thy fame;
>   Thou art indeed, by many a claim,
>   The poet's darling."

Such is the glowing love of that true poet of Nature for this cheerful flower. Burns did not value it less, as his lament, when he was obliged to plough it up, witnesses:—

> "Wee, modest crimson-tipped flower,
>   Thou'st met me in an evil hour;
>   For I maun crush amang the stoure
>   Thy slender stem;
>   To spare thee now is past my power,
>   Thou bonnie gem.
>
> "Cauld blew the bitter biting north
>   Upon thy early, humble birth;
>   Yet cheerfully thou glinted forth
>   Amid the storm;
>   Scarce reared above the parent earth
>   Thy tender form.
>
> "The flaunting flowers our gardens yield,
>   High sheltering woods and wa's maun shield;
>   But thou, beneath the random bield
>   O' clod or stone,
>   Adorns the histie stibble-field
>   Unseen, alone.
>
> "There in thy scanty mantle clad,
>   Thy snawie bosom sunward spread
>   Thou lift'st thy una-suming head
>   In humble guise;
>   But now the share uprears thy bed,
>   And low thou lies!"

The Ox-eye Daisy (Chrysánthemum leucánthemum), is nearly

as common in the meadows and corn fields as our little favourite in the pasture. Its golden disk and silver rays make a brilliant show among tangled grasses, or the red Poppies of the corn field. The beautiful Chrysanthemums, the autumnal ornaments of the gardens of the south and of greenhouses in the north, are the own sisters of our poor Ox-eye.

The Yellow Ox-eye, or Corn Marigold, is less common, though growing in great abundance in some districts. I was delighted once to find a few plants in a corn field in Wiltshire, but last year I could have gathered a whole stack of it. Fields and waste ground were golden with its blooms about Callander, Oban, and in the Isle of Arran. The Germans call it "Gold Blume," and its Latin name means *gold* (C. ségetum, *Plate XI.*, *fig.* 9). Shakspeare honours it with his notice:—

> "The Marigold that goes to bed wi' the sun,
> And with him rises weeping."

And Keats is diffuse in its praises:—

> "Open afresh your starry folds,
> Ye ardent Marigolds!
> Dry up the moisture from your golden lids;
> For great Apollo bids
> That in these days your praises should be sung
> On many harps which he has lately strung;
> And when again your dewiness he kisses,
> Tell him I have you in my world of blisses."

In Denmark these Marigolds are such a nuisance, that a law is made compelling the farmers to eradicate them.

The common Feverfew (Pyréthrum parthénium), grows in waste places frequently. The flowers have yellow centres and white rays. They are arranged in a corymb, and the leaves are powdery. The scent of the whole plant is aromatic.

The Corn Feverfew is equally common. Its leaves are of a bright green, pinnate, with thread-shaped segments. Its flower is large and solitary. My specimens of both are from the neighbourhood of Richmond.

The Sea Feverfew has a reddish tinge in the centre of the

flower; in other respects it resembles the Corn Feverfew. Fanny brought this specimen from Clevedon (P. maritima, Plate XI., fig. 10).

I found the Wild Chamomile (Matricária chamomilla), near Penzance, in July last. The leaves are small and pinnate, and the whole plant prostrate. This plant is cultivated for the sake of its flowers, which are infused for a tonic. We used to be invited to walk about on our Chamomile-bed, in furtherance of the axiom of the old distich:—

> "The more you tread it
> The more you spread it."

The common Chamomile (Ánthemis nóbilis), I found about Looe. It is a much larger plant than the Wild Chamomile.

The Stinking Chamomile (A. cótula), is a common weed in corn fields. It is easily recognised by its unpleasant odour.

There are a Sea Chamomile (A. maritima), with cream-coloured rays; a Corn Chamomile (A. arvénsis), slightly hoary; and an Ox-eye Chamomile (A. tinctória), flowering in corymbs.

The Yarrows are the last family in this great Composite order.

The Sneezewort Yarrow (Achilléa ptármica), grows on road-sides in many places. I have it from Yorkshire, Lancashire, and Argyleshire, and Edward has it from Kent. It has narrow, whitish leaves, sharply cut, and large white flowers.

The common Yarrow (A. millefólium), grows everywhere; on hedgebanks, pastures, and even lawns. The leaves are dark green, and very much divided, and the stem is exceedingly tough. It has large crowded clusters of very small flowers, varying from white to pink.

The Serrated Yarrow (A. serráta), is sulphur-coloured. It grows eighteen inches high. It is an inhabitant of Derbyshire.

The Woolly Yarrow (A. tomentósa), has branched corymbs and bright yellow flowers, pinnate leaves, and a tough stem. It grows a span high, and frequents the Scotch and Irish mountains.

Interesting as the whole of this extensive order is, the

Thistle group is the one most suggestive in its characteristics. It is a constant memento of the sin of our first parents, on account of which fell the curse, "Thorns also and *Thistles* shall it bring forth." Amaziah, in his foolish temerity in thrusting a quarrel upon Jehoash, is contemptuously compared to a Thistle: "The Thistle that was in Lebanon sent to the Cedar that was in Lebanon, saying, Give thy daughter to my son to wife; and there passed by a wild beast that was in Lebanon, and trod down the Thistle." As at once a sign of idle husbandry, and a type of idleness itself the Thistle is disliked on all sides. The children's poet points a warning with it:—

> "I passed by his garden, and saw the wild Briar;
> The Thorn and the Thistle grew stronger and higher."

Nor is the floating seed of the Thistle without its voice of instruction. Ever ready to fall on unoccupied ground, and fill with evil that which before was only guilty of emptiness, it is a type of sin and evil practice, as Mr. Howitt teaches us:—

> "Lightly soars the Thistledown;
> Lightly doth it float;
> Lightly seeds of care are sown,
> Little do we note.
>
> "Lightly floats the Thistledown;
> Far and wide it flies;
> By the faintest zephyrs blown
> Through the summer skies,
>
> "Watch life's Thistles bud and blow,
> Oh 'tis pleasant folly!
> But when all our paths they sow,
> Then comes melancholy!"

CHAPTER XII.

CAMPANULÁCEÆ—LOBELIÁCEÆ—VACCINÁCEÆ—ERICÁCEÆ — MONOTROPÁCEÆ — AQUIFOLIÁCEÆ — OLEÁCEÆ—APOCYNÁCEÆ—GENTIANÁCEÆ—POLEMONIÁCEÆ—CONVOLVULÁCEÆ—BORAGINÁCEÆ.

> " Your voiceless lips, O flowers! are living preachers,
> Each cup a pulpit, every leaf a book,
> Supplying to my fancy numerous teachers
> From loveliest nook.
>
> " 'Neath cloistered boughs, each floral bell that ringeth,
> And tolls its perfume on the passing air,
> Makes Sabbath in the fields, and ever ringeth
> A call to prayer.
>
> " Were I, O God! in churchless lands remaining,
> Far from all voice of teachers and divines,
> My soul would find, in flowers of thy ordaining,
> Priests, sermons, shrines!"
>
> HORACE SMITH.

SOME of the orders now coming before us contain a number of the most beautiful and attractive of our native plants.

The BELL-FLOWER order has five stamens, the campanulate corolla being cut into five segments. The calyx is five-lobed, and remains fastened to the ovary while the seed ripens. The seeds are numerous.

This Spreading Bell-Flower (Campánula pátula), is one of the more rare members of the family. It grows in a straggling manner, reaching a height of two feet. Its stem is branched and angular, and the bell is so open as to verge on the wheel-shape. This specimen grew on a hedgebank between Sutton and Warminster, in Wiltshire. The whole plant is rough.

The Rampion Bell-Flower (C. rapúnculus), is an elegant plant, with a tapering, compact panicle, and smaller purple flowers.

It is pretty frequent in gardens. I found it also near Warminster. This is the true Rampion. Before salad herbs were generally cultivated in England, the root of this plant was eaten raw, and sold for that purpose in the markets.

The corn fields of the same neighbourhood furnished the Corn Bell-Flower, or Venus's Looking-glass (C. hýbrida). The bloom of this little plant is well known in gardens; it is very open, and of a brilliant violet hue. The segments of the calyx are very long, and curl backwards. The leaves are a long oval, and curly.

The Heathbell, or Harebell (C. rotundifólia), is a general favourite, and found everywhere in meadows, pastures, and roadsides, and on wild commons, rocks, and old walls. Though common all over England, it is principally associated with Scotland, being called the " Bluebell of Scotland : "

"My ain Bluebell, my bonnie Bluebell."

"It springeth on the heath,
 The forest tree beneath,
Like to some elfin dweller of the wild ;
 Light as a breeze astir,
 Stemmed with the gossamer,
Soft as the blue eyes of a poet's child.

"The very flower to take
 Into the heart, and make
The cherished memory of all pleasant places ;
 Name but the light Harebell,
 And straight is pictured well
What ere of fallen state lie lonely traces.

"We vision wild sea-rocks,
 Where hang its clustering locks,
Waving at dizzy height o'er ocean's brink ;
 The hermit's scooped cell ;
 The forest's sylvan well,
Where the poor wounded hart comes down to drink.

"We vision moors far spread,
 Where blooms the Heather red,
And hunters, with their dogs, lie down at noon ;
 Some shepherd-boys, who keep
 On mountain sides their sheep,
Cheating the time with flowers and fancies' boon.

> " Light Harebell, there thou art,
>   Making a lovely part
> Of all the splendour of the days gone by,
>   Waving, if but a breeze
>   Pant through the distant trees,
> That on the hill-top grow broad-branched and high.
>
> " Oh! when I look at thee,
>   In thy fair symmetry,
> And look on other flowers as fair beside,
>   My sense is gratitude,
>   That God has been thus good,
> To scatter flowers like common blessings wide."

Smaller than this, and far more fragile, is the little Ivy-leaved Bell-Flower (C. hederæfólia), which Fanny found growing along with the Pale Butterwort, in the swampy ground near the Loe Pool. The little bells are very pale, and the stem has none of the wiry texture of the Harebell; on the contrary, it is soft and tender, and can only rise to the height of a few inches by supporting itself on the stems of other plants. We have a garden species, a size between this and the Harebell, of a pure white. It is a native of France, and is called there "La Religieuse des Champs."

The Creeping Bell-Flower (C. rapunculoídes), is a handsome plant, with a one-sided spike of bright purple drooping flowers and roughish leaves; those on the stem being lance-shaped, those on the root heart-shaped. It is rare as a wild plant; my specimen is from a garden.

The Clustered Bell-Flower (C. glomeráta, *Plate XII., fig.* 1), grows abundantly in fields adjoining Studley Park, near Ripon. The deep purple bells are situated in a sessile cluster at the top of the stem, and one or two more flowers spring from the axils of the next pair of leaves. In these meadows it grows a foot and a half high; but I have gathered the plant upon the Wiltshire downs, and in Switzerland, where the height was only two inches.

The Giant Bell-Flower (C. latifólia), is the glory of our northern woods. Rising to the hight of four or five feet, with a spike of flowers more than half the length of the whole plant, it has a most stately appearance. In Yorkshire the flowers are generally pale lilac, or white, but they vary to purple in other districts. Further south than York the plant is rare.

The Nettle-leaved Bell-Flower (C. trachélium), is also a handsome species, with rough, sharply-serrated leaves, and abundant large purple flowers on its branched stem. The leaves are lance-shaped, inclining more or less to heart-shape at the base. This plant is rare in the north, but abundant in the midland and southern counties. I have found it in profusion both in Warwickshire and Shropshire. It grows also in considerable quantities about Canterbury, and has hence got the name of "Canterbury-bell," though the species really entitled to that name is a garden flower. Clare speaks of a boyish trick played with the flowers of this plant and the glow-worm, by putting the insect within one of the bells, which it illumined, and so forming a mimic lantern. Howitt states that the same experiment used to be made with the Rose.

> "When glow-worm, found in lanes remote,
> Is murdered for his shining coat,
> And put in flowers that Nature weaves
> With hollow shape and silken leaves,
> Such as the Canterbury-bell
> Serving for lamp or lantern well."

The Round-headed Rampion (Phyteúma orbiculáris), is in my hands. It in no way resembles the old-fashioned Rampion, nor do I hear that it was ever put to any useful purpose. This is a rare plant, frequenting chalky downs. The flowers are arranged in a head almost like Clover, and the colour is dark blue or purple. My specimen is from Mr. Ward's garden.

There is a species with elongated clusters of greenish flowers, called the Spiked Rampion (P. spicátum). Sir J. E. Smith gives Mayfield and Waldron, Essex, as its habitat. I have not met with a specimen.

The flowers of the Sheep's Scabious (Jasíone montána), the third family of the Bell-Flower tribe, are of a sky blue. From the florets being arranged in heads, this plant resembles a

Scabious, but its growth is weaker. Its anthers are united, and all the florets are alike. The sepals are large. This plant was gathered in the sandy lanes about Goosetery, in Cheshire. The whole plant is very hairy, and flowers in July.

The LOBÉLIA family comes next. They have biting qualities, which in one species become poisonous.

The Water Lobelia (Lobélia dortmánna), was sent to me from the Cumberland lakes, where the roots, with their clusters of leaves, grow far under water, while the delicate spikes of pale blue flowers rise above the surface.

The other species, the Acrid Lobelia (L. úrens), is peculiar to this county, Devonshire. A kind friend has given me a specimen from near Axminster. The flowers are sky blue, and the leaves lance-shaped and serrated. This is the poisonous plant; it flowers in September.

There are many foreign species of Lobelia of every shade of blue and crimson. Many of these grow at the Cape, and some are domesticated in our gardens.

The poisonous Lobelias are succeeded by the wholesome CRANBERRY order, and my county is rich in these. They are shrubby plants, with eight stamens and one style, the corolla being either wheel or pitcher-shaped. How vividly the day when I collected these specimens returns to my mind! The weather was brilliant even for July, and we joined Mr. ——'s pupils and friends in an excursion to Brimham Rocks. These wonderful masses of mountain limstone are scattered at random over a high moor of great extent, some half dozen miles from Harrogate. From the top of one of them York-Minster can be clearly discerned, and every landmark within twenty miles besides. It was a luxurious day for a botanist. The boys were most obliging, their knives ready at any moment to take up roots or sever branches, and any number of willing hands eager to carry the basket of specimens. The older pupils were even more useful, with their ready classical derivations, and skilful rendering of quantities.

The Bilberry blossoms were fast disappearing, though a delicate little rosy urn remained here and there. Many a Bilberry-gathering have I had on that moor; one, in particular I remember, when our party had picked at least a quart of the berries, and an unlucky fall of the basket-bearer dissipated all our store! The berries are black and sweet, and make a tolerably good tart. The leaves are shiny and serrated, and the stems angular. (Vaccínium myrtíllus).

We found the Cowberry (V. vitis-idǽa), that day, and my young friends rejoiced over it as much as I did. It varies from the Bilberry in having its flowers in clusters, the berries red, and the leaves evergreen and turned back at the edges.

The Bog Whortleberry (V. uliginosum), with its clustering stalks and black berries, we sought in vain.

The elegant Cranberry plant (V. oxycóccus, *Plate XII., fig.* 2), the tiniest of evergreens, was there in full beauty, its pink lily-shaped flowers raising their frail heads above the peat moss. The fruit is much esteemed for tarts, and a great quantity of it is imported every year from Russia and North America. The produce of the plants on our own moors is very small, partly because the unenclosed lands are so largely on the decrease, and partly because the Cranberries are not produced in sufficient quantities to pay people for going far to collect them. The habit of the plant is trailing, and the stems are very slender.

The same excursion was fruitful also in the HEATH tribe, the one succeeding the Cranberry. The tribe takes its name from its first family.

The Cross-leaved Heath (Erica tétralix, *Plate XII., fig.* 3), was there, and here and there one of its rose-tinted bells was open. The flowers grow in tufts at the end of the stem, and the leaves are four in a whorl.

The Fine-leaved Heath (E. cinérea), grew by its side, its purple flowers arranged in whorls along with the leaves. Its blossoms were more fully open than those of the Cross-leaved species, though August is the legitimate month for the flowering of both. This is the species which the poet depicts as singing:—

> "Where the wild bee comes with a murmuring song,
> Pilfering sweets as he roams along,
>   I uprear my purple bell:
> Listening to the free-born eagle's cry,
> Marking the heath-cock's glancing eye,
>   On the mountain side I dwell."

Besides these plants were Ferns and Reeds of various kinds, so that my basket became heavy enough to test the kindness of my courteous friends.

The Cornish Heath (E. vágans), is one of Fanny's trophies from her beloved Lizard district. It carries its flowers in whorls; they are of a pink hue, and the brown stamens protruding through the opening of the corolla give great expression to the plant. The barren tract of land from Helstone to the Lizard Point is covered with this Heath, as thoroughly as the land about Brimham Rocks was with the Cross-leaved and Purple species. The specimens were gathered late in August.

She ought also to have found the Fringed Heath (E. ciliáris). It has flowers in the axils of the leaves, and these clusters hang to one side. I have seen a specimen, but could not beg it.

The Mediterranean Heath is only found in Ireland. I fancy it is one of Professor Forbes' twelve Spanish plants.

I suppose that the Heaths so beloved in Scotland are the Cross-leaved and Purple ones, as they are prevalent in the north of England. The Highlanders thatch with them, and the Picts used to make ale of the young shoots. A heap of Heather is accounted a bed comfortable enough for a Highland shepherd. Carrington alludes to this fact:—

> "How many a vagrant wing-light waves around,
> Thy purple bells, Erica! 'Tis from thee
> The hermit birds, that love the desert, find
> Shelter and food; nor these alone delight
> In the fresh Heath. Thy gallant mountaineers,
> Auld Scotia, smile to see it spread immense
> O'er their uncultured hills; and at the close
> Of the keen boreal day, the undaunted race
> Contented on the rude Erica sink
> To healing sleep."

The one British member of the Ling family has fully as much share as the Erica in the useful purposes we have mentioned. Our peasants about the moor districts thatch with Ling and Heath together, make brooms of the Ling, and what we call "pan-scrubbers"—that is, a little tight bundle cut stiff and square at both ends, which is a very useful scullery article. Ling is also used by them for fuel, and the grouse feed upon it (Callúna vulgáris). It is a shrubby plant, with very small leaves, and numbers of tiny, bell-shaped, purple flowers, with crimson calices. It blooms in August.

The third family in the Heath tribe is the Menziesia, of which there are two members.

The Scotch Menziesia (Menziésia cœrúlea), has its purple blooms in a cluster. The flowers are larger than those of the Heaths. Sir J. E. Smith speaks of it as a rare inhabitant of Perthshire, and recent explorers declare it to have been entirely removed by an Edinburgh nurseryman.

The Irish Menziesia, or St. Dabeoc's Heath (M. polifólia), is a very elegant plant. The large crimson pitcher-shaped blossoms hang separately from different sides of the stem. It grows on some of the Irish mountains, but my specimen is from a garden.

The Trailing Azalea (Azálea procúmbens), is a Scotch plant, but I have not got a native specimen. Mine was gathered on the Niesen, a mountain near Thun, in Switzerland. It is a creeping plant, with woody stems, much branched, and dwarfish. The leaves are opposite, and rolled back at the edges, and the flowers are reddish. The number of stamens in this family is five.

The Marsh Andromeda (A. polifólia), is also a prostrate bush, with narrow alternate leaves, lined with white, and clusters of elegant bell-shaped flowers. My specimen was given me by Mr. Ward, of Richmond.

The Bearberry family is the last in the Heath tribe.

The Red Bearberry (Árbutus uva-úrsi), I have only seen in the Botanical Gardens of Edinburgh. The stems are woody, interlacing, and procumbent; the leaves oval, and the clusters of pinkish, bottle-shaped flowers clothe the end of the branches. It flowers in May and June. The berries are red.

The Alpine Bearberry has black fruit and serrated leaves. I have never seen a living plant, only a portrait of it.

The Strawberry Tree (A. únedo, *fig.* 4), is very handsome, bearing its wax-like blooms and crimson fruit at the same time. A kind friend brought me beautiful specimens from the banks of the Lakes of Killarney, where it grows very abundantly. It is a favourite shrub in our gardens. Martin Tupper speaks of "the waxen flower of the Arbute, which dieth in a day." alluding to the perishable nature of the delicate blossoms. The Latin name "Unedo" means "One I eat," because, as Sir J. E. Smith explains it, when you have eaten one of the beautiful-looking berries you will desire no more.

The Bird's-nest tribe succeeds the Heath tribe. It contains two families, the Winter-green and the Bird's-nest or Fir-rape.

The largest of the Winter-greens grows near Harrogate, the Round-leaved Winter-green (Pýrola rotundifólia). Its leaves are glossy, and ascend on footstalks from the roots. The flowers are white, having five sepals, and a corolla in five segments. The blossoms are hung on alternate sides of the stalk, and look like round drooping bells. The spike is nearly a foot high, and there are bracts at each flower-stalk.

The Intermediate Winter-green (P. média), is smaller than the Round-leaved species, and its flowers are pink. It grows in abundance in woods in Swaledale, flowering in August.

The Lesser Winter-green (P. minor), closely resembles the last-named, but is smaller. I have no specimen of it.

The Serrated Winter-green (P. secúnda), has cut leaves and greenish-white flowers. I have found it abundantly in Switzerland, but never in Britain.

The Single-flowered Winter-green (P. uniflóra), is easily distinguished by its larger solitary flowers. According to Sir J. E. Smith, both these species belong to the hills of Scotland.

This curious unnatural flower, which looks as though it had been buried and dug up again, grows parisitically on the roots of Pine and Fir, and hence it is called Fir-rape; its other name, Bird's-nest, being in reference to the overlapping scales at the foot of the stem, which seems to make a snug nest. The whole plant is of a light yellow brown, and the spikes, bent at first, have somewhat the appearance of horses' heads. The terminal flower has ten stamens, the others have eight. The blossoms are rather bell-shaped, and there is no nearer approach to leaves than the rude scales which adorn the stem. I found my plants in a wooded hollow on Warminster Down, facetiously called the "Frying-pan" (Monótropa hypópitys, *fig.* 10).

Our common HOLLY (Ílex aquifolia), is the only British member of its order (Aquifoliaceæ). Its name is a corruption of *holy*, because it is used to decorate churches at Christmas. Its Latin name means "needle-leaved." They say that the bark has tonic properties; bird-lime is made from it, and the wood is used in inlaying. It is a remarkable fact, that the leaves upon the higher branches are often without spines.

> "Below a circling fence its leaves are seen
>   Wrinkled and keen,
> No grazing cattle through their prickly round
>   Can reach to wound;
> But as they grow where nothing is to fear,
> Smooth and unarmed the pointless leaves appear."
>
> SOUTHEY.

Mr. Johns relates that a certain John di Castro, having learned the method of boiling alum at Constantinople, returned to his own country to pursue his researches in natural history. He found near Tolfa the Holly tree growing; and, as he had observed the same shrub to flourish in the alum districts of Asia, he began to search for alum beneath the soil. Ere long he was able to establish profitable alum works. The first English alum works were opened in the neighbourhood of Guisborough, in Yorkshire. Sir J. Challoner first observed that the foliage in that district was of a very light green, and this suggested to him the presence of alum. The seeds of the Holly are contained in a berry, the scarlet beauty of which I need not describe to you.

> "When I see the Holly berries,
> I can think I hear
> Merry chimes and carols sweet
> Ringing in my ear.
> Christmas with its blazing fires
> And happy hearths I see;
> Oh! what merry thoughts can cling
> Around the Holly tree!"

The flowers are small, and of a creamy white, tinged with lilac. The stamens and stigmas are four in number, and the corolla is cut into four segments. The richest quantity of bloom I have ever seen was in the neighbourhood of Hawkhurst last May.

The Privet (Ligústrum vulgáre), represents the first British family of the OLIVE order. It is a small shrub, with broad lance-shaped leaves, thick-set panicles of white flowers, and black berries. The corolla is divided into four, the stamens are only two, and there is but one stigma. It flowers in July, and abounds on the borders of the Wiltshire downs. We found among its leaves a huge green caterpillar, with beautiful purple bands across it, and a horn upon its tail. The creature had a way of raising itself on its last few pairs of hind legs, and lifting the front part of the body erect in a manner that looked quite threatening. We were afterwards told that it was the caterpillar of the large Privet moth.

I have the uninteresting bloom of the Ash (Fráxinus excélsior), consisting only of brown stamens and stigma, without either calyx or corolla, and opening while the branches are yet naked, so that they have no relief from foliage. The "keys," as country people call the bunch of long seed-vessels, form a far more interesting object, accompanied as they are with the graceful pinnate leaves. White, in his "History of Selborne," gives an account of a cruel ceremony in which a pollard Ash was a principal instrument. It was generally believed that if a shrew mouse touched the limbs of cattle it gave them pains and lameness. The remedy for this was to split an Ash tree, make a hole in the wood, put a live mouse therein, and bind up the tree again. Happily the poor mouse would quickly die of suffocation. Twigs from such a tree, called a Shrew Ash, would cure the cattle suffering from shrew malice. It was also a custom to pass diseased children through a cloven Ash. The shade of the Ash is said to be detrimental both to corn and grass. The Lancashire farmers sometimes give the bark to their cattle. Evelyn says that the keys used to be pickled for salad, and they certainly were used medicinally by old herbalists. The wood is valuable for its toughness; kitchen tables made of it do not splinter. Milk-pails are formed of it, by rolling the plank into a hollow cylinder, and putting in a bottom. Ash timber will bear a greater weight than any other. When burnt, it makes good potash, and the bark is used in dyeing. Homer speaks of the heroes as armed with Ashen spears. The Romans called it Fraxinus; it was much used by them for implements both of war and agriculture. Pliny declares that serpents are terrified of it, and Dioscorides, the physician, prescribes it as a cure for their bites. In the west of England they burn Ash faggots at Christmas, in memory of Alfred the Great's first campaign after his sojourn in the peasant's cot, when he and his soldiers kept themselves warm by burning the Ash trees in their vicinity.

We must not leave the Olive tribe without noticing some of its foreign members.

The Myrtle, the cherished ornament of our greenhouses, forms the common hedgerows at the Cape and in the countries of the east; it grows in great profusion in the Madeiras, and attains the height of a hundred feet in Australia. As introduced in Scripture, it becomes an emblem of repose: the angel, in Zechariah, stood among the Myrtle trees as he said, "We have walked to and fro in the earth, and behold all the earth sitteth still and is at rest." The Jews still observe the command for the observance of the Feast of Tabernacles, and "go forth into the Mount and fetch Olive branches, and Pine branches, and Myrtle branches, and Palm branches, and branches of thick trees." Anne Pratt quotes from Mr. Lane an Arabic tradition, that "Adam fell down from Paradise with three things: the Myrtle, which is the chief of sweet-scented flowers; an ear of Wheat, which is the chief of all kinds of food; and pressed Dates, which are the chief of fruits." She also relates that the Roman ladies used to bathe beneath the Myrtle trees on the 1st of April, and proceed, crowned with the leaves, to offer sacrifice to Venus.

The fragrant Jasmine also belongs to this tribe.

The Olive is the most important family of the tribe. Our garden Lilacs and Syringas belong to it, but not upon these is founded its value. The Olive tree of Palestine and the east furnishes a valuable oil which is used in medicine, in soap-making, and for other purposes of commerce. Its value was very early recognised, for the Sacred Historian puts self-appreciating words into the tree's mouth, twelve hundred years before Christ. "The Olive tree said unto them, Should I leave my fatness, wherewith by me they honour God and man, and go and be promoted over the trees?" The oil drawn from the fruit of the Olive was used both in the service of the sanctuary, for the solemn anointing of kings and priests, and also in social intercourse to anoint guests and brighten the countenance. "Wine that maketh glad the heart of man, and oil to make his face to shine." Two Olive trees spoken of in the fourth chapter of Zechariah, are supposed to be typical of Christ Jesus and the Holy Spirit of God. An Olive branch was the sign given to Noah that God's anger was turned away, and the waters of the flood assuaged; and ever since that leaf has been regarded as an emblem of peace. When the curse shall at last be removed, and renewed creation be restored to perfect beauty, then God will plant "in the wilderness the Cedar, the Shittah tree, and the Myrtle, and the Oil tree." At the present day, the Olive is extensively grown in Italy and the southern parts of Spain and France. The oil is rubbed upon the skin in hot countries.

The Dogbane order succeeds that of the Olive; the members of this tribe are, to say the least, unwholesome. The deadly poison, Nux vomica, is the fruit of one species, and from it strychnine is obtained. Although administered in homœopathic doses their power becomes curative instead of hurtful, yet we cannot repress an involuntary shudder at their names of evil omen.

The Tanghinia Nut, the terrible ordeal poison of Madagascar, is another member of this group. In that dark land a person accused, however falsely, of being bewitched, is compelled to swallow first some portions of the skin of a chicken, and then a quantity of this poisonous plant. Its quality being violently emetic, the unfortunate wretch, after enduring hours of agony, is attacked with violent vomiting. If all the pieces of skin be found the accused is pronounced innocent. But this is rarely the case; and if it is, the patient generally dies from the poison.

There is a very touching account recorded of the virulent poison of the beautiful Oleander, another member of this group. It is related that in 1809, when some French troops were marauding, they cut some branches from the Oleander for spits and skewers wherewith to roast the cattle they had taken. The poisonous juice of the branch entered the meat, and of twelve men who ate seven died, and the other five became dangerously ill.

Yet there are wholesome members of the tribe, and the Cow Tree is one of them. The qualities of this tree are so wonderful and beneficent, that I have copied Humboldt's description of it. "On the barren flank of a rock grows a tree with dry and leather-like leaves; its large, woody roots can scarcely penetrate the stony soil. Its branches appear dead and dried, yet as soon as the trunk is pierced there flows from it a sweet and nourishing milk. It is at sunrise that this vegetable fountain is most abundant. The natives are then to be seen hastening from all quarters, furnished with large bowls to receive the milk, which grows yellow and thickens on its surface. The milk is glutinous, free from all acrimony, and of an agreeable and balmy smell. It was offered to us in the shell of a Calabash. The negroes and free labourers drink it, dipping in their Maize or Cassava bread."

Though our British Periwinkles are acrid, they are innocent herbs when compared to many of their brethren.

The Lesser Periwinkle (Vínca mínor), grows abundantly about Hawkhurst, covering the deep hedgebanks with a carpet of evergreen, studded with purple stars. They have five stamens, and the wheel-shaped corolla is cut into five segments. The Latin name is from a word signifying *to bind;* and the Italians bind down the sods over their tombs with it, on this account calling it "the Flower of Death." Hurdis has some pretty lines on the plant:—

> " See where the sky-blue Periwinkle chinks,
> E'en to the cottage eaves, and hides the walls,
> And dairy lattice, with a thousand eyes
> Pentagonally formed, to mock the skill
> Of proud geometers."

I also have specimens of the Lesser Periwinkle. They come from Sowley Wood, a part of the ancient forest of Selwood; and I have it also from the Grantley Woods, Yorkshire.

I have found the Greater Periwinkle (V. májor), wild near Penzance; it was growing in a hedge far away from any garden, so I think it must have been quite wild. It also grows in hedges near Ross, in Herefordshire. Its leaves are broader in proportion than those of the last-named species, and the plant is in every respect much larger. The bright, full hue of the foliage has earned for it the name of "Little Laurel." The pistil of both the Periwinkles is beautifully fringed.

It is quite a relief after hearing of the dreadful poisons we have discussed to turn to the wholesome GENTIANS. Bitter as is the flavour of every species of this beautiful tribe, all are wholesome and some beneficial. Generally the plants have five stamens, and the corolla is five-cleft. They inhabit both cold and warm climates. The Gentian family have blue or lilac flowers, I mean the British species of it.

The Marsh Gentian (Gentiána pneumonánthe), has bright blue flowers shaded towards the centre of the bell with black and yellow. It has narrow leaves, and at Leckby Scar it reaches a span high; but on Rudd Heath, in Cheshire, where I have also gathered it, the height is not more than two or three inches.

The Spring Gentian (G. vérna, *Plate XII.*, *fig.* 6), with its intensely blue corolla cut into round segments and wheel-shaped, favours our county. Its principal habitat is the neighbourhood of the High Force, Teesdale, where my specimen grew. In size it is smaller than the least plant of the Marsh Gentian.

The Autumnal Gentian (G. amarélla), is a taller plant, at least six inches high. It has a five-cleft corolla, of a clear lilac colour. We found it in some fields on the hills between Wensleydale and Swaledale, crossing by the Butter-tub Pass. These so-called Butter-tubs are strange freaks of Nature: deep holes in the solid mountain with rough rocks jutting out from their sides draped with Ferns. In some the water can be heard to gurgle far below, but in others neither water nor any other bottom can be discerned. The calling of the district being essentially a dairy one, these frightful crevices are named "Butter-tubs." Doubtless water was the principal agent in their formation, washing away the moveable earth and leaving the rocky masses in naked grandeur. The blossoms of this Gentian are arranged in panicles, the side branches also bearing flowers. It blooms in August.

The Field Gentian (G. campéstris), resembles the Autumn Gentian, but is smaller; the colour of the corolla is less clear, and is four-cleft. The plant grows freely on gravelly pastures. We used to gather it on a limestone waste, called Quarry

Moor, and near Wicliffe in the Ripon neighbourhood. I have found it also in abundance on high pastures about Richmond, and in the Yorkshire dales.

The Dwarf Gentian is the noble blue bell which adorns our flower-borders. It grows on the mountains in South Wales, but my specimen is from a garden. I fancy that this is the species taking part in the imagined dialogue of Montgomery :—

> "Blue thou art, intensely blue!
> Flower, whence came thy dazzling hue?
> When I opened first mine eye,
> Upward glancing to the sky,
> Straightway from the firmament
> Was the sapphire brilliance sent."

The Alpine Gentian (G. alpína), has a very small blossom, but the five-cleft corolla is of a brilliant blue. It graces some of the Scotch mountains, but my plant did not grow in Britain. I found it at the foot of the Gemini Pass, studding the sward as with stars of Lapis lazuli.

The Yellow Gentian, from the root of which the valuable tonic drug is drawn, is found in the valleys of Switzerland and Germany; and we used to gather a Bog Gentian there also, with upright clusters of intense blue bells standing on the summit of the stem, and in the axils of the leaves.

The Centaury family is the second in the Gentian tribe. I remember being much amused when in Switzerland by an English gentleman of our party, who, returning from a distant excursion, told me that he had found a plant far exceeding in beauty all that I had yet collected. After this announcement he produced from within his hat a parcel enveloped in his pocket-handkerchief which, when opened, divulged—the common Centaury! But whether common or not, it is certainly a flower of great beauty. The sturdy plant grows five or six inches high; the stem is mostly simple, and the panicle branches proceed from the axils of the upper leaves. The corolla is bright rose colour, divided into five segments, the tube tinged with yellow. It grows on some pasture lands called Ripon

Parks, and in woods about Richmond. All the Centauries partake of the bitter quality of the tribe, and this species has been used with success as a tonic (Erythræa centáurium).

Edward says that Kent furnishes abundant specimens of the common Centaury, and there the little Branched Centaury also grows. Hawkhurst is near enough to the sea to catch the sea breezes, and even a dash of the briny spray in a high wind. This accounts for the Branched Centaury growing there, as it does not approve of inland situations. It is only two or three inches high, shrubby in its growth, and its numerous branches are beset with solitary flowers of a deeper hue than in the last species (E. pulchélla, *Plate XII., fig.* 7).

Fanny brought the Dwarf Centaury (E. littorális), from near Fowey in Cornwall, where it grew near the seashore. The leaves are large and oval; the plant is dumpy, and the flowers pretty, as in the other species, but few.

The Broad-leaved Centaury is by some believed to be only a variety of the Dwarf Centaury. As its name indicates, its leaves are broader, but its habit is much the same. The shores of Lancashire are given by Sir J. E. Smith as its habitat. All the Centauries close their blossoms at noon.

A rare British member of the Gentianella family she found upon raised ground by the side of the Marazion Marsh. That once busy town, Marazion, is now a waste; the earnest, money-making inhabitants mingle with the dust; the mines of tin, by means of which they carried on a trade with the Phœnicians, are closed and forgotten; and the turrets of St. Michael's Mount behold only swamp and heath, where the once prosperous town flourished; and in the place of the Jewish traders rise scarce and beautiful plants of every form and hue, from the stately Royal Fern, emulating a tree in size, to the tiny Gentianella, with its thread-shaped stem of two inches, and its yellow funnel-formed corolla, cloven into four segments (Exácum filifórme).

I have the Yellowwort (Chlóra perfoliáta, *Plate XII., fig.* 8), our only member of its family. The plant grows from twelve to eighteen inches high; the leaves are glaucous, and the pairs being united the stem seems to pierce its way through them: hence it is called "perfoliate." The flower is large and cupped, divided into eight segments, and of a brilliant yellow colour. There are eight stamens and one stigma. It grows about Studley, near Ripon, and I have also gathered it in the neighbourhood of Goostery, Cheshire.

The Buckbean family comes next, and, like the two that precede it, it has only one British representative.

The common Buckbean (Menyánthes trifoliáta, *fig.* 9), is a beautiful plant with trefoil leaves, and a spike of pale, rose-tinted flowers. It has five stamens and one pistil, and the five-cleft corolla is thickly covered on the inside by a white fringe; the leaves have large sheaths at their base, and both these and the stem are tinged with crimson. It is a water plant, and adorns ponds in many places. I have found it about Ripon, and in peat bogs among the Yorkshire hills; also in Cheshire.

The Water Villársia is a very beautiful plant. Like the Buckbean its corolla is fringed; but the flower is cupped, yellow, and of a large size. The leaves are heart-shaped and wavy, and the flower-stalks spring from their axils; the flowers are solitary. It grows in ponds near Ely, and was sent to me from thence. The only time I have seen it was in a ditch near Oxford; but the ditch was broad and deep, and baffled all my attempts to reach its treasures. This is the last family of the Gentian tribe.

The Phlox tribe succeeds that of the Gentians, and our gardens offer many beautiful representatives; but we have no indigenous ones.

The Jacob's Ladder family has but one British member.

The Blue Jacob's Ladder (Polemónium cœrúleum, *Plate XII.*, *fig.* 10), is a very rare plant, and I am proud to exhibit a specimen from Yorkshire. This piece was gathered at Leckby

Scar, a famous botanical haunt near Topcliff. We also have it from Malham Cove, another Yorkshire habitat. The large blue flowers are arranged in a panicle; the corolla is five-cleft, and the stamens five; the leaves are lance-shaped and pinnate. It is a common flower in gardens, and attains a greater height there than when growing wild.

The Convolvulus tribe is the next upon the list. It contains two families—that of the Convolvulus and that of the Dodder.

The common Small Convolvulus (Convólvulus arvénsis), is found everywhere, and is everywhere worthy of admiration. It twines up the stalks of corn, round any erect plant; and, devoid of support, it will creep along a hedgebank, or even over a heap of stones. Its flowers are rose-coloured, and very perishable; its leaves arrow-shaped.

The Great Bindweed (C. sépium), is a magnificent climber, hanging its large bells of pure white in every hedgerow. Its leaves are large, of a full green, and heart-shaped. This plant has many of the properties of its brother, the Scammony, which is very much used in medicine. Jalap is also the produce of a member of this family.

There is a plant of this group, the roots of which are used in the same manner as Potatoes by the Spanish and Portuguese. It was formerly cultivated in England, and called "Sweet Potato;" it is the root spoken of by Shakespeare and other writers of that period.

The Sea Convolvulus (C. soldanélla), the last British member of this family, is a coast plant, one of those which Fanny found growing on the sands at Penzance. It has flowers as large as the Great Bindweed; but the corolla is angular, not round, and beautifully tinged with pink. The leaves are kidney-shaped, and the stem is procumbent. The plant is small. Before she could get the flowers home they faded, and she had no opportunity of visiting the shores again, as they were proceeding immediately to Helston. However, she found it abundantly on the bar of the Loe Pool, and laid specimens at once into her sketch-book.

On the cliffs near Polperro she found the Lesser Dodder twining its crimson threads round Furze bushes. Its flowers grow in clusters; they are like waxy bells. The seed of these plants does not split into lobes, but it opens, and puts forth a little spiral body, which is the embryo.

There is a Greater Dodder (Cúscuta européa), which I once found entwining round a large Onion in our garden; it differs little from the Lesser except in size, and in having an apology for a flower-stalk to some of the florets of the cluster.

There are a Flax Dodder and Trefoil Dodder, parasitic on the plants after which they are named. I have specimens given by Mr. Perry, of Warwick, but it is difficult to discern any points of distinction. All these plants are very noxious among crops, choking the herbs around which they twine. I think our Lesser Dodder was innocent of injuring the Furze (C. epíthymum, *Plate XII., fig.* 12).

We now come to an extensive and very beautiful tribe—that of the Borage. Blue is the prevailing colour of the flowers. The corolla is five-cleft, and has often teeth in the centre; the leaves are generally rough.

The Viper's Bugloss (Échium vulgáre), is one representative of the first family in the Borage tribe. It has a spotted hairy stem, and one-sided cluster of funnel-shaped blue flowers. It is a beautiful plant, and looks brilliant as it lifts its head from the ruins of Richmond Castle, where it grows side by side with the wild Wallflower. I have found it also upon the walls of Jervaulx Abbey, and upon rocks in that district. The spotted stem suggests a likeness to the viper, which seems to be acknowledged throughout Europe. In Spain it is called "Herba de la Vibara," and in France "La Viperine." Our forefathers, pursuing the theory of "Signatures," believed that it would heal the bite of a viper. Gerarde was of opinion that the mere sight of it would affect serpents. He says, "Its vertues are so forcible, that the herbe only thrown before the scorpeon, or any other venemous beast, causeth them to be without force or strength to hurt; insomuch that they cannot move or stir until it be taken away."

There is a Violet-flowered Viper's Bugloss (E. violáceum), but it is peculiar to sandy ground in Jersey, and we have no specimen.

The Lungwort (Pulmonária officinális), with its blotched ovate leaves, is very familiar as a shrubbery plant; it grows wild in woods, but is very rare. The flowers are pink when first they open, but they turn gradually blue. The plant is about a foot high. My specimen is from a garden.

The Narrow-leaved Lungwort (P. angustifólia), is still more rare. They both flower in May.

The Gromwell family comes next. Its seeds are hard, and contain a considerable portion of flint. Its English name is derived from Celtic words signifying *stone seed*. The Latin name has precisely the same meaning. The Germans call it "Steinsame," and the Dutch "Steenzaad." The French name is the prettiest—"Plante aux perles."

The Corn Gromwell (Lithospérmum arvénse), has narrow blunt leaves of a dull green, with white flowers. Its height is about a foot. I have gathered it in corn fields in Yorkshire, Kent, and Wiltshire.

The common Gromwell (L. officinále), grows about Clevedon. The plant is twice the size of the Corn Gromwell, but the flowers are no larger. The corolla is five-cleft, and of a sulphur colour.

The Purple or Creeping Gromwell (L. purpureo-cœrúleum), is a very handsome species. Its barren stems are prostrate and rooting, the flowering ones erect; the blossoms are large, clustered, and of a full purple hue. It also grows within a few miles of Clevedon.

The Sea Gromwell (L. marítimum), resembles this, but is widely spreading, and the hue of the flowers less decided. We have no specimen of it.

The common Alkanet (Anchúsa officinális), falsely so called, Fanny found on the Clevedon sands. It is so very rare a plant that some people cannot believe that it was really wild there; but as no garden was near, we take the benefit of the doubt, and have faith in its being an orthodox native. The spikes are terminal and one-sided, the flowers blue, and the leaves lance-shaped and very rough.

The Evergreen Alkanet (A. sempervírens, *Plate XII., fig.* 14), Edward found in Shropshire, and rejoiced, believing it to be a garden plant. He was not far wrong, for it is much cultivated in old-fashioned gardens, and its brilliant flowers entitle it to notice. It is rare as a wild plant, though found pretty plentifully in Shropshire and Warwickshire.

The common Comfrey (Sýmphytum officinále, *fig.* 11), the next family of the Borages, is very abundant in Wiltshire. The plant grows three feet high, and the tube-shaped flowers are arranged in forked drooping clusters. Near Heystesbury the flowers are white, but about Brixton Deverill they are lilac. The leaves are large, and of a broad lance-shape. The root is used by the villagers as a medicine.

The Tuberous Comfrey (S. tuberósum), has yellowish flowers, and is a slighter plant. It is rare except in Scotland.

The common Borage (Borágo officinális, *Plate XII., fig.* 13), is a very handsome plant. Its drooping clusters of brilliant blue flowers, with their white centre and black stamens, plentifully adorn the cliffs at East Looe, contrasting with the yellow, white, and pink plants which I have already described. It grows elsewhere on rocks and waste places; but I have never seen its beauty so striking as Fanny describes it on those Cornish cliffs. It used to be cultivated for the sake of its flowers, which form an ingredient in the drink called "cool tankard," and it still lingers in old-fashioned gardens. It has been discovered by an eminent chemist that a decoction of the leaves of this plant evaporated deposit crystals both needle-shaped and cube-shaped, the former being of nitre, the latter of sea salt. This circumstance is mentioned in a little monthly volume entitled "Wild Flowers of the Year," and published by the Tract

Society. It further relates that the dried leaves of the plant held near a candle will produce a flash of light.

The Small Bugloss (Lycópsis arvénsis), is a less interesting plant, with smaller leaves, and comparatively inconspicuous blue flowers. The great peculiarity of the blossom is that the tube is bent to one side. It grows on the same cliffs at East Looe, and in fields near Warminster, Wilts.

Now for the favourite family of all—the Scorpion-grasses, or Forget-me-nots.

The prettiest member of the family, the Water Scorpion-grass (Myosótis palústris, *Plate XIII., fig.* 1), is the true Forget-me-not. We never see this lovely flower by the lake or quiet stream without remembering the sad story of the devoted knight, who daring too much to procure the lightest wish of his lady, found the stream too strong for him, and had only power to throw his dearly-bought prize to the shore ere he sank beneath the water. Well was the flower named Forget-me-not, for whenever its turquoise gems are seen, the story of that lover is related. Earle Colne details a pretty dialogue with this flower:—

> " Flower of the modest eye
>   Tell me, who love you well,
> Whence comes your brilliant dye?
>   Why stores your honey-cell?
>
> " Who built your turquoise-cup,
>   Its garlanding of green?
> Who glossed its inner parts
>   With such golden sheen?
>
> Oh! sweet-eyed flow'ret tell me,
>   For my soul doth long to know,
> How you woo our hearts to love you
>   More than flowers of brighter glow."
>
> " There dwells a look in my gentle bloom,
>   That whispers to hearts of the happy home
>   A good man ne'er forgot;
> And because it resembles the kindly eye,
>   Which never met ours but with sympathy,
>   They have called it Forget-me-not."

The flowers grow in curled clusters, destitute of leaves.

The Creeping Water Scorpion-grass (M. répens), varies little from this, except in having leaves in the cluster. I have not found a specimen.

The Wood Scorpion-grass (M. sylvática), is a lovely plant, scarcely inferior in charm to the Water one. It covers our northern woods with starry gems in April and May, growing a foot high. I believe it is a very local plant. Its leaves are smaller than those of the true Forget-me-not, and of a very dull green. The clusters are full and brilliant.

The Field Scorpion-grass (M. arvénsis), is a small plant with tiny sky-blue flowers. It blooms all the summer months, and is common in corn fields, hedgebanks, and waste places.

The Early Field Scorpion-grass (M. collína), grows on Blackford Hill, near Edinburgh. I found it in March; it was only an inch high, and its flowers were proportionately small, and of a very bright blue. Fanny has also specimens of it which she gathered on banks and rocks near Clevedon.

The Yellow and Blue Scorpion-grass (M. versícolor), she found near Polperro. The flowers of this species are also small; but the plant is easily distinguished by having some of the blossoms blue, and the freshly-opened ones yellow. These three last-named Scorpion-grasses are annuals.

The Tufted Scorpion-grass (M. cœspitósa), is perennial. It grows in wet places; Edward has it from Kent, and I from the county of Durham. Its only fault is that there is too much foliage for the small spikes of pale blossoms. The leaves are pale and hairy, and the stems thick, and sometimes a foot high.

The Rock species (M. rupícola), has a very handsome blossom, large, and intensely blue. The stems are tufted. It is found on mountains, but mine is a garden specimen.

The Little Mudwort (Limosélla), is a prostrate plant, bristly, and with crowded leaves, and very small blue flowers. I have sought it in such low pastures as it is said to inhabit, but in vain.

The last family in the Borage tribe is the Hound's-tongue. I have found the common Hound's-tongue (Cynoglóssum officinále), on the road between the town of Richmond and the racecourse. It is a downy plant, with a smell like mice, and dark crimson drooping flowers. The leaves are broadly lanceolate, and the stamens are shorter than the corolla. It grows to the height of a foot and a half. It is abundant in the neighbourhood of Bishopton, in Durham; and I see Fanny has specimens from Looe. This tribe has little except beauty and harmlessness to recommend it. One of the Alkanets is used for a dye, and the root of a species of Gromwell is applied to the same purpose. All the members have insipid juice, and are more or less hairy.

CHAPTER XIII.

SOLANÁCEÆ — OROBANCHÁCEÆ — SCROPHULARIÁCEÆ — LABIÁTÆ — VERBENÁCEÆ — LENTIBULARIÁCEÆ — PRIMULACEÆ — PLUMBAGINÁCEÆ — PLANTAGINÁCEÆ.

> "God might have made the earth bring forth
> Enough for great and small,
> The Oak tree and the Cedar tree,
> Without a flower at all.
>
> "He might have made enough, enough
> For every want of ours,
> For luxury, medicine, and toil,
> And yet not have made flowers.
>
> "Then wherefore, wherefore were they made
> All dyed with rainbow light,
> All fashioned with supremest grace,
> Upspringing day and night?
>
> "To comfort man, to whisper hope
> Whene'er his faith is dim;
> For Whoso careth for the flowers
> Will much more care for him."
>
> MARY HOWITT.

AGAIN we come to a poisonous tribe of plants, the NIGHTSHADES, but there is much to be said in favour of these. They possess a great variety of qualities; some poisonous, some harmless, some actually nutritious. The Deadly Nightshade is nearly related to the wholesome Potato, the mild Tomato to the pungent Capsicum, and the innocent Egg-plant to the poisonous Tobacco. Here, as in many of the Umbellifers, the evil qualities existing in the esculent members of the tribe are dissipated by heat, and thus they are changed from slightly deleterious to decidedly wholesome. Leigh Hunt must have had the medicinal qualities in mind when he wrote—

> "Sage are yet the uses
> Mixed with our sweet juices,

> Whether man or mayfly profit of the balm:
>> As fair fingers heal'd
>> Knights from the olden field,
> We hold cups of mightiest force to give the wildest calm.
>> E'en the terror poison
>> Hath its plea for blooming;
>> Life it gives to reverent lips, though death to the presuming."

The Deadly Nightshade and Henbane are cases in point. Rash meddlers with these get illness, or even death, as the reward of their temerity, while the agonised sufferer blesses the same Henbane for his much-needed sleep, and the painful headache yields before homœopathic doses of Belladonna as to a magic charm. Truly the Nightshades have no need to blush for their members; for besides these beneficent medicines they have the Potato, Tomato, Cayenne Pepper, and Winter Cherry for food, the last being pleasant to the palate when ripened in Spain or Portugal, and the Calabash for drinking vessels. The first family is the one which gives its name to the tribe.

The Woody Nightshade (Solánum dulcamára), is a shrubby plant, climbing to the height of four or five feet. The flowers are in cymes, the corolla divided five times, and the segments turned back; the colour a fine violet. The five stamens grow close together, and appear almost as one; they are bright orange, and contrast perfectly with the violet corolla. The leaves are dark green, smooth, and halbert-shaped, and the berries are bright scarlet. Altogether it is an attractive-looking plant, and grows commonly in hedges. My specimen was gathered near Ripon. The berries and young twigs are slightly poisonous and emetic.

The Black or Garden Nightshade (S. nígrum*), grows frequently as a weed in gardens, and very much frequents waste

---

* The "Apples of Sodom" are the fruit of another member of this family. Murray says, in his Encyclopœdia, "At the foot of the mountains on the shore of the Lake Asphaltites Hasselquist found the Solanum sodomæmum, the fruit of which, internally destroyed by insects, preserves its colour, but contains only dust. These are the 'Apples of Sodom' spoken of by Josephus, and the 'Grapes of Sodom and the Grapes of Gall' of Holy Writ."

places in some localities. About Norwood and Blackheath it puts forth its clusters of small white flowers, and ripens its handsome black berries in great abundance. The flower-stalks spring from the axils of the leaves, which are ovate and bluntly toothed. The whole plant is poisonous, and the berries are so in a high degree.

The Deadly Nightshade (Atrópa belladónna, *Plate XIII.*, *fig.* 2), is a handsome plant, with large, bright green, ovate leaves, and dull purple bells, of an inch long, proceeding from the axils of the leaves. The berries are as large as a Cherry, black, and very glossy. The plant grows between Warminster and Corsley, and in many other places in that neighbourhood. The flowers have a lurid tinge. The berries are very noxious. A detachment of French soldiers, halting near Dresden, were attracted by the inviting-looking fruit, and ate a quantity of it: 180 men were thus poisoned, many of whom died, and the rest were long in recovering. On another occasion eighteen children ate of the berries; they became unable to swallow, and the pupils of the eye grew dilated; they became drunk and furious. They all recovered; but, on going to school four days after, some saw everything as if red, some could not see at all, and none of them could articulate. The Italian ladies use this plant as a cosmetic: hence its name of Belladonna. I know, too, that when administered internally it gives great brilliancy to the eyes, and it is a homœopathic remedy for visual ailments and for sore throats.

I have the pleasure of presenting another poisonous plant, but which is surely too untempting to be dangerous. The Henbane (Hyoscýamus níger, *Plate XIII.*, *fig.* 3), has a powdery appearance, and a very faint disgusting smell. The leaves, which are large and closely set, are lance-shaped and deeply cut, and the dull sulphur-coloured flowers are exquisitely veined with dark purple. They grow in spikes, only two or three compact flowers blooming at once. In bud the spike is curved, but in seed it is tall and straight. The spikes of capsules are a beautiful object when skeletonised and bleached, resembling a large upright Lily of the Valley. There are five stamens and one stigma. Children have been poisoned by eating the seed in mistake for Filberts. I have gathered it near Ripon, in Yorkshire, and Fanny has found it both at Clevedon and in Cornwall.

Sir J. E. Smith has admitted the Thorn Apple (Datúra stramónium, *Plate XIII., fig.* 10), into the British flora. It is a very handsome plant, with large, fragrant, white, trumpet-shaped flowers, and deeply-lobed, bright green leaves. The capsule is fully as beautiful as the flower, covered as it is with strong bristles. When subjected to the process of skeletonising, it presents a most beautiful object, exhibiting a network of fibre, bristling all over with bleached thorns of various lengths. It springs up wherever freshly-imported American shrubs are planted, and sows itself from year to year. In a garden in Wiltshire it has existed thus for a number of years, and the poor people in the neighbouring village come to beg the leaves, which they smoke in the style of a cigar as a cure for asthma. This practice of inhaling the smoke of Stramonium in a paroxysm of asthma is derived from the natives of the East Indies, and I have medical authority for saying that it is a safe means of relief. Notwithstanding this, the leaves have a highly poisonous quality, and are dangerous when swallowed. A friend of mine, an army surgeon, told me that, when with his regiment in Canada, he was called to the wife and children of a soldier, who were insensible, and apparently dying from a stroke of the sun, as it was supposed. The medicine administered acted as an emetic, and they began to recover. On inquiry, it was found that they had been gathering Lamb's-quarter (Chenopódium álbum), to cook for dinner, and had eaten some leaves of the Thorn Apple at the same time. A similar case, equally well authenticated, occurred in New Jersey during the revolutionary war, in which three soldiers fell victims to the Thorn Apple poison; one dying

P

raving mad, and the two others with symptoms of lock-jaw. The best antidote for it is said to be vinegar.

The BROOM-RAPE order comes next; it contains but two families, that of the Broom-rape, and that of the Toothwort. These are parasitical plants, growing on the roots of others. The seeds lie in the ground until they come in contact with some root congenial to them, when they forthwith germinate. They have no leaves, only scales or bracts, and there is no shade of green in any part of the plant.

The first of the Broom-rapes that we shall speak of is the Red species (Orobánche rúbra). In a "Week at the Lizard" it is mentioned that this grows on the Magnesian rocks, between Kynance and the Lizard Point, so Fanny diligently sought it there. Her search was rewarded by a sturdy specimen. The flowers, stem, and scales were the colour of iron rust; the upper lip of the corolla was cloven, and the lower cut into three segments.

Afterwards she found the Tall Broom-rape (O. elátior), in a bushy field near Clevedon. Its stem is whitish, tinged with lilac, and the flowers are all pale, and with lilac stripes and blotches.

Near Warminster the Lesser Broom-rape (O. mínor), grows freely in Clover fields. It is slighter in form than those I have mentioned, and of a uniform brown hue.

The Greater Broom-rape (O. májor), grows upon Furze roots near a field path leading from Horningsham to Maiden Bradley, in Wiltshire; it is tall and stout in habit, and of a purplish-brown colour.

There is a Clove-scented Broom-rape, parasitic on roots of the Hedge Bedstraw and Bramble; Sir J. E. Smith gives Kent as its native county; and there are a Purple and a Branched Broom-rape, both having three bracts to each flower, while the rest have but one. The Purple species requires sea air, whilst the branched one is content wherever it can find roots of Hemp (O. ramósa).

The Toothwort (Lathræa squamária), grows in woods near

Richmond. Its stem and scales are of ivory white, the latter overlap one another, and look like rows of teeth at the base of the stem. The flowers are pendulous, and of a delicate lilac colour; the lower lip is in three segments, the upper lip cloven. My specimen was in bloom in April.

We now come to a pretty numerous order—that of the FIG-WORTS. Many of these have handsome flowers, and some are useful. Our friend the Foxglove (Digitális purpúrea, *Plate XIII., fig.* 4), has both these recommendations. The beauty of its crimson bells none will dispute, and a valuable medicine is procured from it, which, however, needs to be administered by medical care, as its effect is weakening and even paralysing if taken incautiously. Sir Walter Scott pleases to regard it as an emblem of pride, and certainly it has a good right to be vain of its appearance.

> "Foxglove and Nightshade side by side,
> Emblems of punishment and pride,
> Grouped their dark hues with every stain
> The weather-beaten crags retain."

Edward has specimens from Kent, where the plant grows in great luxuriance. Nor does it stint its favours on our Yorkshire moors, all Britain rejoices in its patronage.

The Snapdragon family succeeds that of the Foxglove. My county boasts the Greater Snapdragon (Antirrhinum május, *Plate XIII., fig.* 5), the most showy member of the family. It grows freely on old walls about Richmond, and on the ruins of Jervaulx and Fountains Abbey. The corolla in all this family is of the form called *gaping*; some species have spurs, but this and one other are without them. The flowers are large and gay, of a bright rose colour with yellow between the lips. The leaves are a narrow lance-shape.

The Ivy-leaved Snapdragon, or Mother o'Millions, flourishes on the same old walls, decorating the ruins of Richmond Castle and Easby Abbey, as well as those above mentioned. The leaves are bright dark green, and lined with purple; the small lilac flowers are solitary upon long footstalks.

At Clotherholme, near Ripon, the Yellow Snapdragon (A. linária), grows very freely. Its tall golden spikes adorn the hedgebank, and its closely-set narrow leaves are of a glaucous green. The corolla is shaped like that of the Greater Snapdragon, and is of a light yellow on the lips, while the centre part is of a full orange: hence it is called by country children "Butter and eggs." Our name for it was "Toad Flax." Anne Pratt thus addresses the plant:—

> " And thou, Linaria, mingle in my wreath
> Thy golden dragons; for though perfumed breath
> Escapes not from thy yellow petals, yet
> Glad thoughts bring'st thou of hedgerow foliage, wet
> With tears and dew; lark-warblings and green Ferns
> O'erspanning crystal runnels, where there turns
> And twines the glossy Ivy."

I found a curious variety of the Toad Flax in Wiltshire; the corolla was regularly five-cleft, and had five spurs. It looked like a kind of cone with five hooks at its base. Smith and Sowerby give a drawing of it, and call it "variety Peloria."

I was staying near Warminster in harvest-time, and searching one day in the stubble fields my eye caught a small creeping plant with oval hairy leaves of a pale hue. Gathering it, I found that small flowers were situated upon long footstalks springing from the axils of the leaves. The corolla was gaping, the upper lip chocolate-coloured, the under yellow. It was the rare Round-leaved Snapdragon!

In similar fields, close upon the chalk downs, and on walls, I found the Least Snapdragon (A. mínus). The plant grows about four inches high, with a branched woody stem, narrow leaves, and minute grey flowers.

Near Warminster I had the good fortune to meet with the Lesser Snapdragon (A. oróntium). Its stem is almost simple; its flowers scarce, and without spur; the corolla purplish-rose colour, with a yellow centre; and the segments of the calyx longer than the flower.

The Sharp-pointed Snapdragon (A. elatíne), Edward has

brought us from Kent, where it grows freely in corn fields, climbing up the haulms and twisting round them. Its flower is yellow and chocolate, like those of the Round-leaved species, and the leaves are of the same soft light shade, but arrow-shaped instead of round.

There is a Creeping Snapdragon with bright lilac flowers striped with blue, and sweet-scented; but we have no specimen of it. It is very rare.

We come next to the true Figwort family.

The Knotty-rooted Figwort (Scrophulária nodósa), grows freely in the Kentish lanes; both it and Water Figwort (S. aquática), flourish by the side of a road called "Hart and Oak." The former has heart-shaped pointed leaves, and a square stem with decided edges; the latter blunt leaves, oblong, and somewhat heart-shaped, and the edges of the stem winged. Both have crimson lips to the corolla, and greenish tubes. They bear their flowers in panicles, are smooth herbs, and smell disagreeably. They are considered unwholesome, and cattle seldom eat them. Anne Pratt relates that when Rochelle was besieged by Cardinal Richelieu, the soldiers were in such distress from famine that they were obliged to resort to the Water Figwort, which abounded there, for sustenance. The French have called it "Herbe du Siège" ever since.

The Balm-leaved Figwort is a downy plant, and its flower is paler. Sir J. Smith says it is very rare, so we need not be discouraged at not having found it.

Yorkshire furnishes us with the Yellow Figwort (S. vernális). I gathered it in April on a piece of waste ground at the entrance of a pasture field near Patrick Brompton, in Yorkshire. It is a light green plant, with crumpled, broad-shouldered leaves, and small panicles of bright yellow flowers, reminding one of a small Calceolaria.

The Cow-wheat family comes next. Their calyces are tubular, containing two seeds.

The Yellow Cow-wheat (Melampýrum praténse), is common

in our woods, flowering in June and July. The leaves grow in pairs, and are deeply toothed at the base; the pale yellow flowers, also growing in pairs, are turned to one side. The corolla is closed, so that you cannot see into its throat or tube.

There are a Crested Cow-wheat with square stems, and heart-shaped, toothed bracts; and a Purple Cow-wheat with pinnate bracts. The flowers of both these are purple.

The Wood Cow-wheat (M. sylváticum), has very small, bright yellow, gaping flowers, and narrow simple leaves. It grows in Teesdale.

The Yellow Rattle (Rhinánthus crista-gálli), is a common meadow plant. The rattling of the ripe seeds in the large capsule is the origin of its familiar name. The upper lip of the corolla is drawn over the style. The flower is yellow, with two or three blue spots.

The Dwarf Red Rattle, or Lousewort, is an exceedingly pretty plant. Both it and the Marsh Lousewort have swelled seed-vessels with pinnate sepals attached. The leaves are finely pinnate, and the rose-coloured blossom rises above the dark sepals. The Dwarf species (Pediculáris sylvática), grows only two or three inches high; it favours moist, elevated pastures. The Marsh Lousewort is from a span to a foot high, with crimson blossoms and a branched shrubby stem. It is comparatively rare: I found it in an excursion up Swaledale. A party of us were walking and riding by turns, and we halted to eat our luncheon on the borders of a wood. There were lead mines very near. During the halt I roamed along a path leading by the side of a swamp, hoping to find some floral treasures. Some half dozen yards off in the wet ground grew the Marsh Lousewort (P. palústris), and the large Cotton Grass side by side. Both were new to me, and I was meditating how best to reach them, when two miners came by. They stopped and regarded me with good-natured curiosity. "You're not ganging into 't bog wi' them bits o' shoas of yer feet?" asked the elder man. "Yes," I replied, "I have a great wish for those two flowers, and I had rather get my feet wet than not have them." "Stand by," said the first speaker, "and this young chap 'll get 'em for yer; his shoas are mair likely for such an a bog as that than yours." The young man did gather the flowers for me, and replied to my thanks with a short nod, and the words, "Ye're welcome." Yet many persons have a great dread of these miners, and account them as savages. True, they do not deal in flattering titles; they tell you that "the parson's a tidy large chap," and the squire's lady is a "real nice woman." But they have honest kindly hearts, and you might wander alone for fifty miles, and never meet with rudeness.

Now look at the Bartsia family. This little dingy plant with the one-sided spikes of dull lilac flowers is the Red Bartsia. It has lance-shaped serrated leaves, and a square stem, growing about a span high. There is a red tinge over the whole plant.

I have the Viscid Bartsia (Bártsia viscósa), a much prettier plant. It has larger leaves of a delicate green, and yellow flowers growing from the axils. The whole plant is sticky, and strongly scented: indeed, its general aspect and odour closely resemble the Musk plant. Fanny brought the specimen from Marazion Marsh.

There is an Alpine Bartsia with somewhat the same habit, and red flowers; but I have not seen it.

The family of the pretty little Eyebright comes next. Our one British member of it is a cheerful little plant, growing freely among the scanty herbage of downs and hilly pastures. The flowers spring from the axils of the leaves, which are ovate, furrowed, and sharply toothed. The corolla is white, variegated with yellow and purple. Old herbalists speak of the plant as a remedy for diseases of the eye, and Milton evinces a similar conviction when describing the Archangel's interview with Adam:—

> "Then purged with Euphrasy and Rue
> The visual nerve, for he had much to see."

I used to imagine that the idea of the efficacy of our pretty Eyebright (Euphrásia officinális), was past for ever, along with the prejudices in favour of Lungwort, and Saxifrage granules; but I find that homœopathic physicians still prescribe Euphrasia for ailing eyes, and I was rejoiced to see a friend's visual organs strengthen and brighten under the influence of minute doses of the pretty plant.

The Cornish Moneywort (Sibthórpia európǽa), is a small trailing plant with round hairy leaves, closely resembling those of the Marsh Pennywort or White-rot; its flowers are solitary, and pale pink.

The next family is a very attractive one. The Speedwells have a wheel-shaped corolla instead of a gaping one, which is four-cleft; they have two stamens and one stigma. Some of the family bear their flowers on terminal spikes, some on lateral spikes or clusters, and some have the flowers solitary.

On the cliffs near Clevedon Fanny found one of the scarcer species—the Spiked Speedwell (Verónica spicáta). Its stem is simple, its leaves getting narrower as they ascend the spike. The flowers are deep blue, and closely packed together. The plant is about a span high, and the narrow spike occupies a third of the stem.

There is a Welsh Speedwell (V. hýbrida), with a thicker spike, but so nearly resembling this that it can hardly be regarded as a separate species.

The Shrubby Speedwell has pinkish blossoms veined with lilac; the cluster is terminal, and many-flowered; but it is seldom that many blossoms open at once, so that it looks less gay than the Spiked species. It lives among the Scotch mountains. My specimen is from a garden plant.

The Rock Speedwell (V. saxátilis), was given me along with the Shrubby Speedwell; it grew by its side on the rockwork of the florist. It is also an inhabitant of the Highland rocks, and its spare cluster of large brilliant blue flowers with crimson centres must be a great ornament to them.

There is an Alpine Speedwell (V. alpína), with paler blue flowers and glaucous leaves, but I have not met with it.

Only one of the Speedwells with terminal clusters is now wanting. The Smooth Speedwell (V. serpyllifólia), raises its humble clusters of little grey flowers, interspersed with ovate shiny leaves, in pastures everywhere. Those species mentioned already bloom in July and August; this, their humble relative, flowers a month earlier.

The Brooklime (V. beccabúnga), is a water Speedwell, growing in ditches. Its clusters are lateral—that is, they spring from the sides of the stem. The flowers are very small, but of a very dark brilliant blue, and the clusters are freely bestowed on every stem. The leaves are ovate and shiny, and the whole plant is thick and succulent.

The Water Speedwell (V. anagállis), has long acute leaves and longer clusters. The flowers are larger, and generally of a lilac colour, though they vary to blue and white. It grows by running streams. Both these species are common in Yorkshire and elsewhere.

We found the Narrow-leaved Speedwell (V. scutelláta), first in a bog on Grinton Moor, in Swaledale. It is a delicately-formed graceful plant, contrasting strangely with the strong growth of the two other water species. The leaves are grass-shaped, and tinged with crimson; the stem weak, and the two or three clusters sparingly adorned with pale lilac-veined flowers on separate footstalks, placed at right angles with one another.

The common Speedwell (V. officinális), is a pretty compact plant with rough, ovate, serrated leaves. The bright lilac flowers are arranged in spikes which grow from the axils of the leaves. It flowers in May, and is common in hilly pastures and hedgerows. The leaves are sometimes made into tea, and the French call this plant "Thé de l'Europe."

The Germander Speedwell (V. chamœdrys), is the species most familiarly known. Its lateral spikes of bright blue flowers are very attractive in spring, and a plot of this plant in full bloom is a very lovely object. A touching anecdote is told of the philosopher Rousseau relating to this plant. In his earliest and happiest days he was exploring the country round Geneva in company with a trusted friend. They came upon a plot of the Germander Speedwell, and paused to admire its cheerful loveliness. Thirty years afterwards he again visited that neighbourhood. In the meanwhile his name had become the watchword of learning and philosophy "falsely so called," and though he had attained fame, he was a stranger to happiness. He stood gazing on the beautiful prospect, his inward eye scanning gloomily the view of his strange and unsatisfying past; suddenly his eye fell on the identical plot of Germander Speedwell which had delighted him so much when his heart had been light and innocent; the associations it awakened were too painful for his self-command, and he burst into tears. Thus does it sometimes please God to address the conscience of man by the voice of a humble flower. Children often call this plant "Bird's-eye," and the corn-law poet evidently associated it with that name.

> "Blue Eyebright, loveliest flower of all that grow
> In flower-lov'd England! Flower whose hedgeside gaze
> Is like an infant's! What heart does not know
> Thee, clustered smilar of the banks, where plays
> The sunbeam on the emerald snake, and strays
> The dazzling rill, companion of the road."

The Mountain Germander Speedwell (V. montána), closely resembles the one just mentioned, but its blooms are lilac, its spike less crowded, and the tint of the serrated leaves is paler. It grows freely in hilly woods, and is very abundant about Richmond, in Yorkshire.

The Green Procumbent Speedwell (V. agréstis), is common on waste ground, and as a weed in gardens. Its flowers are solitary, their footstalks longer than the heart-shaped cut leaves. The corolla is grey.

The Blue Procumbent Speedwell (V. pólita), has brighter flowers; it grows in similar situations.

Buxbaum's Speedwell (V. Buxbáumii, *Plate XIII.*, *fig.* 6), closely resembles these, but has very long footstalks to the flowers, which are larger, and of a brilliant blue; and the capsules are heart-shaped, the lobes downward. This is a scarce species, but grows in a pasture called " the west field," near Richmond.

The Wall Speedwell (V. murális), has its flowers nearly sessile; they are small, and of a bright blue. It grows on walls near Easby, and is a little shrubby plant one or two inches high.

The Ivy-leaved Speedwell (V. hederácea), is a trailing species, with cheerful little lilac flowers, and is common, from the beginning of April, on waste ground.

The Blunt-fingered Speedwell has palmate leaves and purple flowers; it is a native of Norfolk.

The Spring Speedwell (V. vérna), has pinnate leaves and pale blue flowers, resembling the Wall Speedwell in habit. We have no specimens of these last two species.

The Mullein family is the last in the Figwort tribe. These plants have a wheel-shaped corolla cut into five segments, five stamens, and one stigma; the stamens are beautifully fringed.

The Great Mullein (Verbáscum thápsus), has large leaves that look as if they were made of flannel. It grows three feet high, and the leaves are planted near to one another all up the stem. The large yellow flowers seem to repose in a nest of flannel, for the calyces are as woolly as the leaves, and the apex of the spike seems enveloped in wool. Country people call it "Flannel plant." In Switzerland it was cultivated in cottage gardens, and we used to observe that, if we passed early in a morning, no flowers were to be seen on its spikes. It was some time before we found out the reason of this absence of bloom. The custom was for the children to pick off all the full-blown flowers while the dew was on them, and the mothers preserved them for plasters for the chest; they were considered to lose their virtue if the dew dried before they were gathered. The plant grows freely in lanes about Ripon.

The Black Mullein (V. nígrum, *Plate XIII., fig.* 7), flourishes most luxuriantly in the vicinity of Patrick Brompton, adorning waste ground with its beautiful spikes. It grows as high as the Great Mullein, but its flowers are much smaller, and of a paler colour. To atone for this, they are placed on the spike in clusters, so that the tall stem is thickly crowded with them. The leaves are large and pointed, of a full green, and very deeply veined. The violet stamens give great expression to the bloom; the anthers are orange.

Near Crakehall I found a beautiful White Mullein, also with violet-fringed stamens. The flowers were large, and the spike thinly scattered with them. The stem was taller and less firm than in the other two species. There was no garden near, and yet I could not be sure that it was a wild plant, for it did not answer the description of any of the Mulleins. Fanny found a similar plant on the sands at Clevedon; and when we showed our specimen to an eminent botanist, we were told that they were from a rare white variety of the Moth Mullein (V. Blattária).

Fanny found a plant resembling this on hilly pastures near Looe, with yellow flowers; this was the true Moth Mullein. The buds are very pretty, for the calyces are edged with brown starry hairs.

There are a Large-flowered Mullein, six feet high, with the flowers scattered at some distance from each other up the stem; a White Mullein, with downy leaves and cream-coloured flowers; and a Hoary Mullein, very woolly, and with scarlet anthers and a panicled stem. All these are very rare, and favour chalky soil chiefly.

The large order of the Lipped Plants or LABIATES succeeds that of the Figworts; they are all wholesome. Perfume is distilled from some of them, as Lavender-water, Patchouli, &c., and Mr. Johns informs us that some of them are ingredients in Eau de Cologne. They have their culinary virtues also, Sage, Marjoram, Thyme, Basil, and Mint being the "pot-herbs" used in soups, sausages, &c. Two species of Mint are used medicinally, and from one is drawn the essence which flavours those delights of children—"mint-drops." The botanist Bentham has given much study to this tribe, and he asserts that the members only frequent warm and temperate climates: 1000 species belong to the eastern hemisphere, 600 to Europe, and only 159 to tropical countries, even these growing on mountains. The aroma is contained in glands placed in the leaves. Most of the members of this order are herbaceous plants, but the Rosemary of our gardens attains to the dignity of a shrub. A decoction of Rosemary forms an excellent wash for hair, making it curl well; it is often put into hair-grease. It improves soiled black garments to sponge them with Rosemary tea; this is the Labiate plant which Mr. Johns mentions as forming an ingredient in the Eau de Cologne. In Yorkshire we have an old superstition, that where the Rosemary flourishes the wife is master in the house; but where the Rosemary pines in the garden, then the husband is the master. The poor people always carry sprigs of this tree to funerals, and throw them into the grave. It is the largest plant of the Labiate tribe. In Crete, according to Sieber, the regular firing-woods are Sage, Thyme, Cypress, Marjoram, and Lavender. Thus the Labiates minister to the comfort of man in an unusual manner, and not only is the summer air scented with their fresh odour, but the very smoke is perfumed with their fragrance.

The Gipseywort (Lýcopus európæus), represents the first British family of the tribe. I have found it near Leamington, and Edward has it from Hawkhurst, in Kent. It grows in wet places, and gipsies are supposed to darken their complexions by washing with an infusion of it. It has the family characteristic of a square stem and rough opposite leaves. Its flowers grow in whorls, and are very inconspicuous. The corolla is four-cleft, and it has only two stamens.

The Sálvia family is the next in order. We have two British Salvias.

The Meadow Clary is a handsome plant, cultivated for the sake of its flowers. It is found apparently wild near Cobham, in Kent. It is used to flavour the wine called by its name. We used to find it plentifully in Switzerland, and its rich purple spikes of hooded flowers were splendid enough to contest the prize for beauty with our garden Salvias.

Another handsome Sage, a frequent inhabitant of Swiss meadows, the Sálvia glutinósa, is there gathered and spread under beds, and is thus believed to destroy or warn away certain unwelcome insects which abound in those romantic-looking Swiss châlets.

The Wild Clary (Sálvia verbenáca, *Plate XIII.*, *fig.* 8), has a leaf like Sage. Its flowers are also of a rich blue and very handsome; but it is seldom that more than two or three are open at once, so that the spike is rarely showy. It grows on the steep bank upon which Richmond Castle stands, and flowers from June to September.

The Sage, which we associate with roast duck and goose, is a member of this family. Like the Gipseywort, they have only two stamens.

The Mint family succeeds that of the Sage, and its members are numerous.

The Spear Mint (Méntha víridis), is the plant used for lamb sauce; it is a straight-growing plant with bright green lance-shaped leaves, and spikes of small pinkish-lilac flowers. It grows in marshy ground near Birk Park, Swaledale.

The Pepper Mint is used for distilling. The leaves are broader, and there is a camphor-like scent about the plant.

The Penny Royal creeps on the ground; its flower-stalk and calyx are woolly. It is of a much smaller habit than the rest of the family. My specimen was sent from the Pentland Hills.

The one that most nearly resembles it is the Corn Mint (M. arvénsis). Its stem is partly prostrate, its flowers have a crimson hue, and it frequents wet corn fields. It is very common.

The Hairy Mint (M. hirsúta), is, perhaps, the handsomest of the family. Its pale lilac flowers grow both in heads and whorls, its leaves are ovate, and the whole plant is downy. It grows pretty commonly. My piece was gathered near a well in Wensleydale.

The Horse Mint has a shaggy spike and pointed leaves, which are woolly beneath.

The Round-leaved Mint (M. rotundifólia), has an acrid smell, and its stamens are very prominent.

The stem of the Tall Red Mint is zigzag and smooth; and that of the Bushy Red Mint nearly so. We have not found specimens of these plants.

But which of us does not "know a bank where the wild Thyme grows?" I know twenty such. Most fully do I sympathise in the love that the ancients had for the smell of this flower. When they wished to praise any author they said he smelt of Thyme. The Greeks considered it as an emblem of activity, and it is certainly an incentive to the practice of it among the bee-folk. Sheep that have browsed where it grows yield the best mutton, and it was considered to improve venison greatly (Thýmus serpýllum). The plant is too familiar to need description.

The Basil Thyme (T. ácinos), is a pretty plant; recumbent like the Wild Thyme, but with stiff stems, and bluish-grey flowers, beautifully spotted with purple. Like the Mints, this family have four stamens, two long and two short; and scented leaves. Those of the Basil Thyme are broad and pointed. I found it in chalk stubble fields in Wiltshire.

The Calamint I have not found; but I have a specimen of the Lesser Calamint (Calamínta nepéta), which it so closely resembles as to be considered by Sir J. E. Smith hardly a separate species. It is a shrubby plant, with broad, serrated leaves, and numerous whorls of lilac flowers on the forked stems. It grows on the road between Patrick Brompton and Crakehall.

The Marjoram (Oríganum vulgáre), of kitchen celebrity

comes next. Its Latin name signifies "Joy of the Mountain," and it is supposed that its abundance in the Oregon territory was the occasion of that district being so called. It is a tall, brittle plant, much branched; its crimson whorls are in panicles. It grows freely on the banks of the Ure, near Ripon.

The common Bugle (Ájuga réptans), is a familiar ornament of our woods and moist meadows. Its upper lip is very short, almost wanting; but its crowded spike of blue flowers, interspersed with purple-tinged leaves, is a pleasing object. The plant is very smooth and succulent.

The Pyramidal Bugle (A. pyramidális), is a much scarcer plant; its spike has more the form of a pyramid. Its tiny upper lip is deeply cloven, and it has no creeping runners. My specimen was sent from Tenby last June.

Smith considers the Alpine Bugle (A. alpína), a variety of the common species. This plant used to be considered as a cure for cuts; and there is a proverb in France to the effect— "He needeth neither physician nor surgeon who hath Bugle and Sanicle."

I have the Wood Germander (Teúcrium scorodónia), a member of the seventh family of the Labiatæ. It grows about Hawkhurst and elsewhere, in all the hedgebanks. The spikes are one-sided, the flowers of a greenish-cream colour, and the leaves a dark sage green.

The Wall Germander (T. scórdium), has pretty whorls of rose-coloured flowers, and notched leaves tapering at the base. I got my specimen off a garden wall.

The Water Germander (T. palústris), is a rare plant with dull purple flowers, growing in pairs, and a procumbent stem.

The Black Horehound (Ballóta nígra), grows along with the Wild Clary on the Castle Hill at Richmond. A decoction of this taken milkwarm is considered an excellent remedy for colds by country people, but I cannot believe in it except as an emetic. It has a most disagreeable smell, ovate leaves and lilac whorls of flowers. Its height is about a foot and a half.

The White Horehound (Marrúbium vulgáre), has pale flowers and white-tinged leaves. It is also used as a village medicine, and grows in waste places.

The Motherwort (Leonúrus cardiaca), I have not been able to find. It has a good deal of white wool about it; its flowers are purple, and its leaves lanceolate, some three-lobed and some simple. The stems grow to the height of two feet in hedges. It is not common.

The Weasel-snout or Yellow Archangel (Galeóbdolon lúteum), is a very old friend of mine. It abounds in a little wood on the west side of Hewick Bridge, Ripon, and we used to go there on holiday afternoons. It has lance-shaped, serrated leaves, large, spotted, yellow flowers, and a square hairy stem. Little children call the clusters of buds nuts, and play with them as such. It is abundant, too, in Kent and Wilts.

The common Hemp Nettle is a familiar weed in corn fields, with its rough leaves, three-branched stem, and cream-coloured flowers.

The Red Hemp Nettle (Galeópsis ladánum), grows well there also. It is likewise three-branched, and its pairs of crimson flowers are prettily spotted with white; the leaves are narrower, and comparatively small, and the plant is of a much slighter habit; the stem is tinged with crimson.

The Downy Hemp Nettle I have never found. It resembles the common Hemp Nettle in its colour, and the Red Hemp Nettle in its habit.

The Large-flowered Hemp Nettle (G. versicolor), I found in Cheshire. It is a good deal like the common species in its manner of growth, but the flower is large, yellow, and with purple blotches on the upper part of the lower lip.

The White Dead Nettle (Lámium álbum), is the most showy member of its family. We call it White Archangel, and forgive it for its resemblance to a Nettle in consideration of its pretty white flowers. Its smell is disagreeable, but good Bishop Mant only thought it negatively pleasant. He says—

> "And there, with whorls encircling grand
> Of white and purple-tinted red,
> The harmless Nettle's helmed head;
> Less apt with fragrance to delight
> The smell than please the curious sight."

I suppose he uses the term "purple-tinted red" with reference to the Red Dead Nettle, which abounds as a weed in gardens.

A fallow field near Leamington supplied me with the Cut-leaved Dead Nettle (L. incísum), which I afterwards found in waste places near Ross; it differs from the Red species in having its leaves more deeply serrated, and in having the stem bare from the root to the top, where leaves and flowers grow in a crowded style; and the Henbit Dead Nettle (L. amplexicáule), which, when it opens its flowers at all, which it seldom does, shows small bright blooms. Its leaves clasp the stem.

The Spotted Dead Nettle (L. maculátum), has crimson flowers prettily spotted, and nearly as large as those of the White Dead Nettle. It is really handsome, and grows about Clevedon and Yatton, in Somersetshire.

The Betony family is the next in order.

There is but one British member of the family, the Wood Betony (Betónica officinális). The flowers are crimson, and the spike is interrupted; the leaves oval and notched. They are very frequently infested by a tiny fungus, which dots all over the back of the leaf like Fern seed. This plant used to be thought highly medicinal—a kind of universal cure: it was a common saying, "May you have as many virtues as Betony." The plant, if eaten fresh, will produce intoxication.

The Woundwort family numbers five species.

The Hedge Woundwort (Stáchys sylvática), is common. The flowers are maroon, six in a whorl. The leaves are rough and heart-shaped. The whole plant has an unpleasant smell; it grows three feet high. I have gathered it frequently in hedges about Ripon and Richmond. It was formerly used for dressing wounds: hence its name.

The Marsh Woundwort (S. palústris), is a Kentish species, and so also is the Corn Woundwort (S. arvénsis); at least, we found them both there. Their names indicate most correctly their particular habitats. The Marsh species is as tall as the Hedge one, with narrow sessile leaves and handsome purple flowers, somewhat more crowded than those of its brother above mentioned. The Corn Woundwort is a much smaller plant, with lilac flowers and heart-shaped blunt leaves.

The Downy Woundwort (S. germánica), is a tall handsome plant; its leaves, stems, and calyces covered with a profusion of white silky hairs, from among which the whorls of lilac flowers, which adorn its long spike, appear. My specimen came from the Botanic Garden at Edinburgh.

The Pale Woundwort (S. ánnua), has yellow flowers; it is peculiar to the neighbourhood of Rochester.

The common Cat Mint (Népeta catária), grows near Richmond. It is a downy plant, with white flowers, spotted with crimson, and arranged in whorls. The principal interest attaching to it arises from the love cats have for it. It is almost impossible to get a young plant of it to grow, because the cats will not let it alone. An old doggerel says of it:—

> " If you set it
> The cats will get it;
> But if you sow it
> The cats wo'n't know it."

Our cheerful little friend the Ground Ivy (Glechóma hederácea, *Plate XIII., fig.* 9), comes next. We used to dislike it as children, because when we were seeking sweet Violets we were often misled by the blossoms of the Ground Ivy, and so we called it "the Deceiver." But its introduction into fairy tales, where it is mentioned as both knowing and showing the way to the enchanted well, completely reconciled us to it. The plant was formerly used as an ingredient in ale.

The Wild Basil (Clinopódium vulgáre), is the one British plant of its family. It has distant whorls of bright crimson flowers, with pairs of soft, unserrated leaves at each whorl. It was formerly planted upon graves, and is associated with sadness and bereavement. It forms a leading feature in Boccaccio's pretty story of "Isabella," as translated into verse by Keats. The heroine buried the head of her murdered lover—

> "And covered it with mould, and o'er it set
> Sweet Basil, which her tears kept ever wet."

The Wild Balm (Melíttis melissophýllum), is a beautiful plant, perhaps the most showy of the tribe. It grows about a foot high, with lance-shaped, serrated leaves, and a hairy stem. From the axils of each pair of the upper leaves spring two handsome blossoms; the corolla is sulphur-coloured or pinkish, and there is one or more large blotches of violet on the under lip. My specimen was sent me from Newton-le-Willows, in Yorkshire. I have it also from the neighbourhood of Lynmouth, in Devon, and Fanny describes it as growing abundantly about woods in Cornwall.

The Prunella or Self-heal (Prunélla vulgáris), is a very common plant, flowering by roadsides and pasture fields. The flowers are arranged in close whorls, forming a spike; but as only one or two flowers in the crowded spike are ever open at once, the plant appears to be either only just coming into bloom, or very nearly over. The leaves are smooth, ovate, and entire; the flowers bright blue, and the bracts mingled among them have a crimson tinge. The plant is low in stature.

The Greater Skullcap (Scutelláaria galericuláta), I got near Reading, from a ditch close to the railway.

I saw the Arrowhead blooming in that ditch from the window of a railway carriage, and, as we had an hour to wait at Reading, I ran back and got the plant. The Greater Scullcap grew beside it on the bank. It height is from a foot to a foot and a half; its has narrow, lance-shaped leaves, each pair on the upper stems having two pretty lilac flowers between them; the calyx is shaped like a low helmet or skullcap: hence its name.

The Lesser Skullcap (S. minor), is not a span high, with entire leaves, heart-shaped at the base. The flowers are in pairs, and of a redder lilac than those of the other species; the lower lip is variegated with white. It is a much scarcer plant than the Greater Skullcap, of which I see Edward has specimens from the banks of the Avon, in Warwickshire. I found my plants on the margin of a little lake called Sheerwater, on Lord Bath's property, in Wiltshire. It grew along with the Bog Pimpernel, adding beauty to a scene already very rich in charms.

The numerously-represented order of Labiates is succeeded by one with but a single British member.

The common Vervain has a tubular calyx and corolla; the flowers are pale lilac, and placed sparingly on a tall tough spike. The lobes of the ovary are all fastened together in this order, which distinguishes it from the Labiates, where the four lobes are free. The Wild Vervain (Verbéna officinális), was one of the plants regarded with superstitious veneration by the Druids; they directed that it should be gathered about the rising of the Great Dog Star, but so as for neither sun nor moon to be above the earth to see it. It was then considered to confer the power of foretelling future events. The Druids reverenced the Vervain nearly as much as the Mistletoe, and introduced it into their religious ceremonies. This is the "holy herb" of Dioscorides, who ascribed great powers to it, especially in incantations; and it used to be much esteemed as an ingredient in the preparation of love philtres. Mr. Morley pretended to cure scrofula, by suspending this plant round the neck of the patient, as recently as the last century. Sir Walter Scott alludes to the prevalent belief in its powers:—

> "Trefoil, Vervain, John's Wort, Dill,
> Hinder witches of their will."

Pliny speaks of its use in telling fortunes, and of the estimation in which the Romans held it. It is a very unattractive plant, with tough stems, spare spikes of small lilac flowers, and powdery-looking, toothed leaves. It is not found at all in Ireland, but is pretty common in England about villages.

The Lemon-scented Verbena is a native of Chili, and was discovered and introduced into England by Dombey.

I have a specimen of the common Vervain from the vicinity of Richmond, and Fanny has one from Clevedon.

The BUTTERWORT order comes next, and contains two families —that of the Butterwort, and that of the Bladderwort.

The common Butterwort (Pinguícula vulgáris, *Plate XIV.*, *fig.* 1), grows in swamps about Richmond and in Swaledale. It has the appearance of a Violet upon a long stalk, with a star of thick, succulent, pale green leaves close to the ground. As soon as the plant is taken from the earth the stem curves round, and all the leaves turn back in a semicircle. It is used to coagulate milk: hence its name.

The Alpine Butterwort (P. alpína), is a rare Scotch plant. Its leaves closely resemble those of the common Butterwort, but its flower is cream colour, and has a tuft of yellow hairs within. I gathered my specimen on the Niesen, Switzerland.

The Large-flowered Butterwort (P. grandiflóra), is an inhabitant of bogs in the south of Ireland, probably one of the plants belonging to the Spanish flora. The blossoms are twice as large as those of the common species, and have blue veins; the corolla is five-cleft. It is a very beautiful plant; the only specimen I have ever seen was brought from Glengary, one of its few Irish habitats.

The Pale Butterwort (P. lusitánica), is a very elegant little plant. Its stem is hairy, and its bloom pale pink. Its foliage resembles that of the other species. Fanny has it from the bog by the Loe Pool, where the Ivy-leaved Bell-Flower grew, and I have found it in abundance on similar ground between Brodick Bay and Corrie, Isle of Arran.

The Greater Bladderwort (Utriculária vulgáris, *Plate XIV.*, *fig.* 2), is Edward's spoil from a deep pond near Hawkhurst.

It grew in the centre of the pond, with just its spike of flowers above the water, and he was obliged to wade more than knee-deep to reach it, his feet sinking in the mud. Like the Butterwort family, this has only two stamens and one style; the calyx has two segments, and the corolla has an upper and an under lip. In the Greater Bladderwort the upper lip is beautifully veined with red, and the calyx, flower-stalks, and bracts are violet. The leaves are thread-shaped, repeatedly branched, and furnished with bladders; these, being filled with air while the plant is in bud, gradually float it to the surface, but when the flowers are over the air leaves the bladders, and water takes its place; then the plant sinks back to the bottom.

In the Intermediate Butterwort (U. intermédia), the upper lip of the corolla is long and flat, and the bladders are separate from the leaves.

The Lesser Butterwort (U. mínor), has a very short spur, only half the length of that of the Greater one, and its flowers are smaller. My specimens of these were sent from Hampshire.

We now come to the PRIMROSE tribe. The corolla in this tribe is generally salver-shape or wheel-shaped. The tribe boast few members of any utility, though many of them are famous for beauty. No family of flowers is more familiarly known than that of the Primrose.

The common Primrose (Prímula vulgáris, *Plate XIV.*, *fig.* 3), is a universal favourite; its cheerful blossoms seem to welcome and rejoice in the return of spring, as Wordsworth says:—

> "Through Primrose tufts in that sweet bower
> The Periwinkle trailed its wreaths;
> And 'tis my faith that every flower
> Enjoys the air it breathes."

And when he wishes to paint the extreme of insensibility, he writes:—

> "The Primrose by the river's brim
> A yellow Primrose was to him,
> And nothing more."

Howitt, too, celebrates its praises, and, with a heart yearning over the weary little children in our large manufactories, he begs Primroses for them:—

"Gather the Primroses,
 Make handfuls of the posies,
Give them to the little girls who are at work in mills:
 Pluck the Violets blue;
 Ah! pluck not a few!
Knowest thou what good thoughts from heaven the Violet instils?

"Ah! come and woo the spring,
 List to the birds that sing,
Pluck the Primroses, pluck the Violets,
 Pluck the Daisies,
 Sing their praises,
(More witching are they than the fays of old);
Come forth and gather them yourselves,
Learn of these gentle flowers whose worth is more than gold."

"Worship the God of Nature in your childhood;
Worship Him at your work with best endeavour;
Worship Him in your sports; worship Him ever;
Worship Him in the wild wood;
Worship Him amidst the flowers;
In the green wood bowers
Pluck the Buttercups, and raise
Your voices in His praise."
       HOWITT.

The blossoms appear in March or April. It abounds in woods and hedgebanks in many parts of Yorkshire, and Edward will tell us that it is just as abundant in Kent.

Its brother the Cowslip (P. véris), is equally frequent with us. We used to have grand excursions to gather Cowslips. There were some fields near Hewick, in the Ripon neighbourhood, which afforded a plentiful harvest of these flowers, and a busy evening it always was after the spoil had been brought home. We used to make Cowslip tea of the petals, adding sugar and lemon-juice. It was such a treat to brew tea for ourselves, that we always sat down to our feast with great delight; but we found it difficult to finish the beverage when we had drunk a little of it. Then we made balls of the Cowslips, by placing about thirty of the clusters astride upon a piece of string, and then tying the ends very tight. But the greatest use of the Cowslip is in making wine. When judiciously managed it is a pleasant liquor, and very useful in infantine disorders. Mixed with the same quantity of water it was a great solace to us during the measles, moistening the parched throat and inducing sleep. In the neighbourhood of Ripon there are old women who make it their profession for the season to collect Cowslips, pick off the corolla, and offer the article for sale. The price for the corollas, or, in technical language, "the pips," is 1s. 6d. per peck. As the French call the Cowslip "Herba de la paralysie," I suppose it must have been used as a medicine for that disease; but from its sedative quality I should imagine it more likely to induce than to cure paralysis. This flower is a great favourite with the poets. Shakspeare makes the fairy sing:—

> "The Cowslips tall her pensioners be,
> In their gold coat specks you see;
> Those be rubies, fairy-favours,
> In those freckles live their favours:
> I must go seek some dewdrops here and there,
> And hang a pearl in every Cowslip's ear."

And Milton speaks of

> "The flowery May, who from her queen lap throws
> The yellow Cowslip and the pale Primrose."

And Clare honours them with an address:—

> "Bowing adorers of the gale,
> Ye Cowslips, delicately pale,
> Upraise your loaded stems;
> Unfold your cups in splendour. Speak!
> Who decked you with that ruddy streak,
> And gilt your golden gems?"

The Oxlip (P. elátior), resembles the Cowslip in form and the Primrose in colour. The flowers grow in a cluster, and are much larger than those of the Cowslip. It is comparatively a rare plant, frequenting bushy places. We used to find Oxlip plants in woods about Clotherholme and Studley, in the Ripon neighbourhood, and in similar situations about Richmond; but we always procured them at the expense of a torn sleeve or wounded hand, on account of the brushwood amongst which they grow

On boggy ground near Kirklington and Richmond, also in Swaledale, we find the beautiful Bird's-eye Primrose (P. farinósa). The flowers are smaller than any of those in its family except the Scotch Primula, and of a beautiful lilac, arranged in full clusters, like a garden Verbena. The leaves are simple, and both they and the stem are covered with white powder.

The Scotch Primrose (P. scótica), is a still smaller plant, and its blossoms are of a deep purple, with yellow centres. Its leaves are finely toothed and powdery, like those of the Bird's-eye Primrose. It grows on the seacoast in the north of Scotland.

Beautiful Yellow and Purple Auriculas were brought to us from the mountains when we were in Switzerland; they also belong to the Primrose order, and are the original of our beautiful garden Auriculas.

The Polyanthus is a member of the same family.

The Water Violet family has but one British member. It is most unfitly named, for there is not the very slightest resemblance to a Violet; it is much more like the Cuckoo-Flower.

Our Water Violet (Hottónia palústris, *Plate XIV., fig.* 4), has thread-shaped, pinnate leaves, all under water. The pink flowers grow in whorls, and have yellow centres; there are five petals and five stamens. Until within the last few years there was a pond near the Bishop of Ripon's Palace which was covered with this plant, but the pond has been filled up. The Water Violet still grows in a pond near the North Bridge, at the back of a cottage called Noah's Ark, because during the frequent floods of the Ure it is often surrounded by water. My specimen came from that pond.

The Cyclamen family are far better known in gardens and conservatories than as wild flowers. We have, however, one native species, and Fanny had the good fortune to find it near Fowey, in Cornwall, and Edward has it from Kent.

The Ivy-leaved Cyclamen (Cýclamen hederæfólium, *Plate XIV., fig.* 5), has a thick root, which, in countries where it abounds, is the favourite food of the wild boar: hence the plant is called "Sow-bread." The leaves are dark green, with a paler stain upon each; they die away as summer advances, to make way for the blooms, which appear in September. The wheel-shaped corolla is split into five segments, which are turned back. I have seen specimens from Suffolk.

The Scarlet Pimpernel (Anagállis arvénsis, *Plate XIV.*, *fig.* 7), is a well-known member of the Primrose order. This very pretty little plant, so familiar in corn fields and waste places, is called "the Shepherd's Weather-glass." It always shuts before rain; but each day it closes at noon, so it must be early in the day when its state can be regarded as a criterion of the weather. It has five stamens, and its wheel-shaped corolla is divided into five segments. Its oval leaves are bright, and beautifully spotted with black underneath. This flower and the Red Poppy are the only examples that we have in Britain of true scarlet flowers.

The Blue Pimpernel (A. cœrúlea), so common in our gardens, is occasionally found wild. I have it from corn fields in Durham; it closely resembles the Scarlet Pimpernel, except in the colour of its flowers.

The Bog Pimpernel (A. tenélla), is a contribution from Cornwall; it forms plots on Marazion Marsh covered with rose-coloured flowers. The petals are less expanded than in the other species, and this gives the blossom more the form of a bell than a star; the tiny leaves are roundish and pointed, and are placed on the short slender stems in pairs; the flower is terminal. When a plot of Bog Pimpernel is in full bloom it represents a rose-coloured cushion. This family of plants are well named Anagallis, or *Laughter-causing*, for their beauty does cause gladness. Every species is pleasing, whether it be the Scarlet Shepherd's Weather-glass, or the deep blue stars of the Purple Pimpernel, or the rosy bells of the tiny Bog-plant—all are perfect in loveliness.

The Loosestrife family come next. Here, as in the last

family, the corolla is wheel-shaped, and both it and the calyx are five-cleft, enclosing five stamens.

The Moneywort (Lysimáchia nummulária), is a very gay plant; its stems prostrate, and creeping along the ground in every direction, thereby covering large spaces. Its leaves are ovate and pointed, of a light glossy green, and grow in pairs. The flower-stalks spring from the axils, and are rather longer than the leaves. The segments of the calyx are large, and those of the corolla not fully expanded, so that the flower partakes slightly of the bell shape; it is large, and when you meet with an entirely-open flower, it is as large as a sovereign. My specimens grew in a moist meadow near Kenilworth, and Edward describes it as growing in profusion in woods and on hedgebanks about Coalport and Ironbridge, in Shropshire. I have occasionally seen it wild in Yorkshire.

The Wood Loosestrife, or Yellow Pimpernel (L. némorum), is a common ornament of moist woods in every county: it greatly resembles the Scarlet Pimpernel, but is rather larger.

The Great Yellow Loosestrife (L. vulgáris, *Plate XIV., fig.* 6), I am proud to introduce to you, for it is really a handsome plant. Disdaining the creeping habit of its fellows, it rears its head erect, reaching the height of two feet and upwards. It bears its flowers in terminal panicles, each blossom being nearly as large as those of the Moneywort, and of a full yellow. The leaves are opposite, or whorled, and of a broad lance-shape. Edward found it growing by the moat at Bodinan Castle, a beautiful ruin not many miles from Hawkhurst. I have it also from the Spa Valley, near Aldfield, in Yorkshire.

The rarest of the Loosestrifes is also a Yorkshire plant; it is called the Tufted Loosestrife (L. thyrsiflóra). Its flowers are smaller than those of the other species; and the segments into which the wheel-shaped corolla is divided are narrow. The leaves, too, are narrow and in pairs, and the numerous flower-clusters spring from their axils. Leckby Scar, near Topcliffe, is its habitat. My specimen came from thence. July is the flowering time for all this family, though the Wood species appears earlier.

The Chaffweed (Centúnculus mínimus), is a humble sturdy little plant with bright pairs of leaves, in the axils of which tiny red flowers are seated.

The Chickweed Winter-green (Trientális europǽa, *Plate XIV., fig.* 8), is a scarce plant in England. It was found in June last by a Ripon botanist at Brimham, and is one of the plants belonging to the Northern Flora of Forbes; the stem is slender, the pointed leaves in a whorl, from whence the cluster of flower-stalks emerges. Each blossom resembles a small Wood Anemone, but it has seven stamens, and the corolla cloven into seven segments. It is the only British plant with seven stamens, and in Linnæus's arrangement forms a class by itself.

Here is the Sea Milkwort (Gláux marítima), which Fanny brought from the salt mud of the Looe River; it is a fleshy little plant with glaucous oval leaves, in the axils of which bright pink or whitish sessile flowers are placed. There are five stamens, and a corolla of five segments. It flowers in June.

She got the Brookweed, too (Sámolus valerándi), from the banks of the Fowey River, near Lostwithiel; and I have a bundle of specimens from Brodick Bay, Arran. It has pale green glossy egg-shaped leaves, and its clusters of small white flowers terminate a stem of six or eight inches high. This is the last family in the Primrose tribe; it has five stamens, and the corolla in five segments.

The THRIFT order comes next, and contains only one British family.

The common Thrift (Státice arméria), is a familiar ornament of sea-cliffs, its narrow fleshy leaves forming close tufts, and its bright pink heads of flowers rising plentifully from them. It is frequently used as an edging for flower-beds; and I remember wondering, as a child, how any one could prefer the short uninteresting Box-edge to the line of gay Thrift. I have also specimens of this plant from the Wensleydale hills, where a sea breeze can hardly ever reach them. These plants have five stamens and five styles; the corolla is deeply cloven, the calyces are funnel-shaped.

There is a Plantain-leaved Thrift (S. plantagínea), peculiar to Jersey, but I have not got a specimen.

The Sea Lavender (S. limónium, *Plate XIV.*, *fig.* 9), grows in abundance on salt marshes near Clevedon, bending its panicles before the tide twice in the day. The leaves are broadly lance-shaped and fleshy, and the flowers resemble those of our garden Lavender. They partake of the nature of the everlasting, for you may keep the clusters many months without their materially changing either form or colour.

The Upright-spiked and the Matted Sea Lavender were sent to me from the coast of Norfolk; their distinctive peculiarities are indicated in their names.

The Plantain tribe is next in succession. Its two families are the Plantain and the Shoreweed.

The Plantain family has four stamens and one style, and the corolla is divided into four segments, which are turned back. The flowers are arranged in dense spikes.

The Sea Plantain (Plantágo marítima), has thick thread-shaped simple leaves, and tall narrow spikes; it abounds on the cliffs about Looe.

The Buck's-horn Plantain (P. corónopus), grows in similar situations; its leaf is toothed or pinnate, and downy.

The Greater Plantain (P. májor), is common on roadsides and waste ground; its large, broad, ribbed leaves are familiar everywhere. Its chief use is for bird seed.

The Hoary Plantain (P. média), has the most claim to beauty; its oval spike of pink flowers is both pretty and fragrant. Its leaves are ovate and downy.

The Ribwort Plantain (P. lanceoláta), or Rib-grass, forms good food for sheep. Its leaves are narrow, and its stamens black. We used to call the spikes "Jack-straws," and many a good game I have had with them fighting my fifty against my neighbour's fifty.

The Shoreweed family, of which we have but one British member (Littorélla lacústris), varies from the Plantain in having its stamens and stigmas on different flowers. The male flowers are conspicuous, rising higher than the Rush-like leaves; they are solitary, and the stamens are long and handsome. The female flower is concealed among the leaves. The plant grows freely in the salt marshes about Clevedon, and on the banks of the tidal rivers in Cornwall.

With this family the third, or corolla, subclass ends. It is a large class, and comprehends some of the most extensive tribes, all of which have the stamens fastened to the corolla, which is undivided or *monopetalous*.

CHAPTER XIV.

CHENOPODIÁCEÆ—POLYGONÁCEÆ—ELEAGNEÆ—THY-MELÁCEÆ—ARISTOLOCHIÁCEÆ—EMPETRÁCEÆ—EUPHORBIÁCEÆ—URTICÁCEÆ.

> "By the silver founts that fall,
> As if to entice the stars at night
> To thy heart; by grass and Rush,
> And little weeds the children pull
> Mistoook for flowers; oh! beautiful
> Art thou, O earth! albeit worse
> Than in heaven is called good!
> Good to us that we may know
> Meekly from thy good to go;
> While the Holy Crying Blood
> Puts in music, kind and low,
> 'Twixt each ears as are not dull
> And thine ancient curse."
>
> Mrs. Barrett Browning.

We have now discussed three of the subclasses of the first great class of plants, the *Two-lobed* (Dicotyledonous). One more subclass still remains, and this is characterised by having no corolla, and called *Petalless*.

The first order in this fourth subclass is that of the Goosefoot. It includes a large number of uninteresting-looking plants, but which are not, however, wholly without their uses. The tribe takes its name from its first family, which has many members, with their calyx divided into five segments, enclosing five stamens and two styles.

The Good-King-Henry, or Mercury Goosefoot (Chenopódium Bonus-Henrícus), is a common weed in waste places. Its young shoots were formerly used as Spinach, and they are still so in France, where it is highly valued, both for the table and for making into poultices. As a proof of their high appreciation of this herb, the French have named it after their favourite king, Henry IV. The leaves are triangular-shaped, and the spike is compound and leafless; the flowers are green. My specimen grew near Healaugh, in Swaledale.

The Upright Goosefoot (C. úrbicum), is equally common, and very much resembles the Mercury Goosefoot, but has larger seeds and a more upright stem.

The Red Goosefoot (C. rúbrum), has its leaves more elongated, and they have a mealiness about them. I have gathered it on waste ground in many counties.

The Sharp-leaved Goosefoot (C. acutifólium), I found near Norwood; its distinctive feature is its oval-pointed leaves.

The Many-spiked species (C. botryódes), is peculiar to the seashore, and has fleshy leaves; and the Nettle-leaved Goosefoot (C. muralis), has powdery flowers and a fetid smell.

The White Goosefoot (C. álbum), is a common weed. It has white powdery leaves, and is found in gardens and on dunghills.

Fanny found the Maple-leaved Goosefoot (C. hýbridum), near Clevedon. It has a strong and disagreeable smell, and its leaves are notched like those of a Maple, or still more in the Oak style. It is a slender plant, and of a light green colour.

The Sea Goosefoot (C. marítimum), is the neatest of the group. It is small, and has awl-shaped glaucous leaves, which mingle in the spike. It grows in the salt marsh near Clevedon.

The Round-leaved Goosefoot (C. polyspérmum), Fanny found also near that place; and the Stinking Goosefoot (C. olidum), lurks in waste places near seaside towns.

Then there are a Fig-leaved and an Oak-leaved species, both of which Sir J. E. Smith describes as growing near London.

The useful garden Spinach is a near ally of the Goosefoots.

The second family of the tribe, the Oraches, have little more to recommend them. They have five stamens and one stigma, and differ from the Goosefoots in having some flowers with stamens only, some with the styles only, and some with the two united, all on the same spike.

On the banks of the Looe River I discovered a large patch of glaucous foliage; the ovate leaves being very thick, I tasted them, and found them salt and juicy. In August the plants put up spikes of green flowers of the kind I have just described, and my curiosity proved to be the Shrubby Orache (Átriplex portulacoídes).

At Clevedon the Halberd-leaved Orache (A. pátula), grows freely, its prostrate stems embracing the sand or shingle, and its leaves being thick and fleshy; growing inland, its habit is more erect and its leaves thinner.

The Frosted Sea Orache (A. laciniáta), is clothed with silver scales, and its leaves are trowel-shaped. I have gathered it on the shores of the Firth of Forth.

The Grass-leaved Orache (A. littorális), inhabits salt marshes, as does also the Stalked Orache, each of the fruitful flowers of which is situated on a footstalk. I have not found these species.

My friend the Sea Orache (A. marítima), makes a good pickle; its stems may be eaten boiled as Asparagus, and the whole herb when cooked forms a far from despicable dish. I cannot pretend that it equals Spinach, but still I feel that the Oraches are a degree less uninteresting than the Goosefoots. If, however, utility is to be the criterion, we must give the preference to the Beet family.

Our British Beet (Béta marítima), grows on rocks by the seacoast; it is very plentiful in Cornwall. The leaves are ovate, wavy, and of a full bright green. The flowers grow in pairs up a long spike; they have five stamens and two stigmas, and are all perfect. It flowers in June and July.

The Beet of our gardens is another member of this family. Its root is a pleasant winter vegetable, and is capable of being made into tasteful ornaments for garnish. Sugar is made from it to a considerable extent in France.

The Mangold Wurtzel is also a plant of this family, and the farmer will bear testimony to its high value as food for cattle.

The Prickly Saltwort I have from the shore at Granton, near Edinburgh, where it grows in great abundance. It is a prostrate plant, and its slightly rose-tinted calyces give the appearance of a starry corolla. The flowers are situated in the axils of the leaves, which end in a sharp spine; three bracts are placed at the foot of the flower. This plant contains a large proportion of soda; and Harvey tells us that the Round-leaved Saltwort (Sálsola satíva), is largely cultivated in southern Europe, and furnishes the best soda used in this part of the world. It is called Spanish soda.

The poet considers our Little Saltwort (S. káli), an emblem of humility:—

> "The Saltwort's starry stalks are thickly sown,
> Like humble worth, unheeded and unknown."

The Glasswort family is the last in the Goosefoot tribe. These plants have jointed fleshy stems and no leaves. The flowers appear at the joints, and consist in a swelled undivided calyx, one stamen, and one style.

The common Glasswort (Salicórnia herbácea, *Plate XIV.*, *fig.* 10), grows freely in the salt marshes at Clevedon, and is sometimes pickled under the name of Samphire, which, however, it does not at all resemble, being bright green and leafless. Cattle like the salt taste of this plant, and it is so beneficial to them that they are often sent to a shore where it grows to regain lost strength. Like the Saltwort, this plant is often burned for the soda it contains, and is thus used in the manufacture of soap and glass.

The Creeping Glasswort (S. rádicans), Edward brought from Sandgate, on the coast of Kent. It has a woody stem.

There is also a Procumbent Glasswort (S. procúmbens), but we have not got it.

The BUCKWHEAT order comes next. Its members are not very attractive, but they have at any rate an advantage over the Goosefoots, both in this particular and in that of utility. Some members of the Buckwheat tribe provide valuable medicine, as Rhubarb and Bistort; some afford food for man, as the stalks of Rhubarb and the leaves of Sorrel, and still more the true Buckwheat (Polygonum fagopyrum), which, though considered merely fit for pheasants and fowls in this country, is made into cakes in Russia, the flavour of which I am assured is super-excellent.

The Knotgrass (P. aviculáre), another member of this group, is very good for sheep. The seeds of this family are triangular, and the calyx is coloured with pink and divided into several segments.

The Amphibious Persicaria (P. amphíbium), has only five stamens; the leaves are heart-shaped, and the conical spike of pink flowers looks very pretty floating upon a pond. I fished my specimen out of the Avon, near Guy's Cliff, where the famous Guy Earl of Warwick lived.

The Spotted Persicaria (P. persicária), has six stamens; its spike and leaves resemble those of the Amphibious species, but are smaller, and the leaves have a dark stain upon them.

The Pale-flowered Persicaria (P. lapathifólium), is distinguished from its Spotted brother by its greenish flowers, and leaves with glandular dots. Both are common in ditches and watery places.

The Biting Persicaria (P. hydrópiper), has also six stamens, its clusters are long and not crowded, and they droop in an elegant curve. The leaves are of a narrow lance-shape. It grows in Kent and about Hastings.

The Small creeping Persicaria (P. mínus), I found by the side of the Isis, not far from the railway station at Oxford. Its leaves are narrow and flat, and its slender clusters are upright. All these Persicarias have two styles, and most of them six stamens.

We now come to a portion of the family which have three styles.

The Bistort, or Snakeweed (P. histórta, *Plate XIV.*, *fig.* 11), grows in pastures and meadows. It has a tall Reed-like stem, with one leaf sessile upon it. The terminal spike of flowers is very handsome. The root-leaves are stalked and ovate. The root forms a valuable medicine in low fever, and an acid very efficacious in stopping bleeding is prepared from it. The blossoms are pale pink, and have a disagreeable smell. A bouquet of them was once very politely presented to me at Fountains Abbey by a very shy gentleman. Not until after he had given the flowers to me did he take some himself, and carry them to his nose; great was his embarrassment when he found the nature of the scent he had unwittingly bestowed.

The Alpine Bistort is the prettiest, though not the most imposing member of the family. Its spikes of white blossoms, only four or five inches high, with grassy root-leaves half that length adorn our Yorkshire hills. High upon the Swaledale Moors, where the Cloudberry drinks moisture from the clouds which rest on the ground, does the little Alpine Bistort flourish in full luxuriance. The seeds often germinate before they leave the calyx: hence its Latin name "Viviparum."

The Knotgrass Persicaria is found on waste ground everywhere. Sheep eat it with relish, and we greet the rosy blossoms, sessile in the axils of the lance-shaped leaves, with pleasure, because any bright thing is so welcome by the trampled road.

The Bindweed Persicaria (P. convólvulus), is the most elegant of the family, twining Convolvulus-like round the stems of corn. Its leaves are heart-shaped, and generally beautifully tinted with crimson. The clusters are lax, and the large seed-vessels have a handsome appearance. It is a common weed in corn fields.

The Buckwheat (P. fagopýrum), has a handsome flower-cluster, and its arrow-shaped leaves are very pretty. Plots of it are grown in the neighbourhood of large game-preserves, and the seed is used to feed pheasants.

The Seaside Knotgrass (P. marítima), closely resembles the common species; its even glossy seeds are its mark of distinction. I have found it on the shore in the Isle of Arran.

The Dock family resembles that of the Persicaria in having its seed triangular-shaped. The calyx is cut into six segments. There are three styles and three stigmas.

The Curled Dock (Rúmex crispus), is a very common species,

its leaves are crumpled, lanceolate-shaped, and pointed. Its pale green flowers hang in numerous drooping clusters. My specimen was gathered on the banks of the Swale, close to a low part of the road called "Standing Waits." I suppose because the frequent floods often made the way impassable, and the traveller must stand and wait.

The Broad-leaved Dock (R. obtusifólius), is equally common, growing in waste ground in the same and other neighbourhoods. It is distinguished by its blunt leaves.

There are two species of this family which bear the name of Sorrel; they differ from the others in having the stamens and stigmas in different flowers, and in having a strongly acid principle in their leaves. When cooked, or mixed with salad, these plants are wholesome; but as they contain a small portion of what is called "salts of lemon, or oxalic acid," a large quantity of them would be actually poisonous.

The common Sorrel (R. acetósa), with its crimson-tinted leaves and crimson seed, is a familiar object in every meadow; its leaves are arrow-shaped. Cooked as Spinach it makes a delicious dish.

The Sheep's Sorrel (R. acetosélla), is common in high pastures; it forms crimson plots on the Richmond race-course, and on the Swaledale moors. Its leaves are less clearly arrow-shaped, but partake of that form.

The Great Water Dock (R. hydrolápathum), is the handsomest of the set; it grows five feet high, and its huge lance-shaped leaves are slightly glaucous and very much waved; and the crowded whorls are nearly leafless. I found it by the Avon, near Guy's Cliff, just opposite the floating Persicaria.

The Yellow Marsh Dock (R. palústris), with its interrupted cluster, I found in the same neighbourhood; and the Sharp Dock I found near Hawkhurst. This is a comparatively slender plant, with small seeds, and elongated heart-shaped leaves.

The Bloody-veined Dock (R. sangúineus), grows also in those Kentish lanes.

The Golden Dock (R. marítimus), with its yellow clusters, does not grace our collection. It is a seaside plant.

Neither have we any specimen of the Fiddle-leaved Dock (R. púlcher).

I cannot leave this family without rendering to them my grateful acknowledgments for Nettle stings cured; for, however scientific men may pooh-pooh the idea, I am certain that the leaves of the Dock have a great effect in removing the smarting from Nettle wounds.

The Mountain Sorrel (Oxýria renifórmis), is a small plant, all green, and with kidney-shaped leaves.

The Oleaster tribe comes next. There is only one British member of it—the Sea Buckthorn: this is a low shrub with silver-lined leaves, green flowers, orange berries, and abundant thorns. I have seen it in the Edinburgh Botanic Gardens.

The foreign families of this tribe are of more importance. Two species which bear brown berries are cultivated in Persia, and their fruit sold under the name of Zinzeyd, and the Orange fruit of another species is eaten in India and China. The fruit of the rest of the tribe is very soft and insipid. The blossoms of the species called "Olivier de Bohème" are exceedingly fragrant.

The Laurel tribe is closely allied to this, but, having no British representative, it has no place in our native botany.

The Camphor, Cinnamon, and Nutmeg, as well as the Daphne of the ancients, belong to this Laurel tribe. This is the Daphne into which, as Ovid describes, Apollo's nymph was transformed.

> "Be thou the prize of honour and renown,
> The deathless poet and the poem crown.
> Thou shalt returning Cæsar's triumph grace,
> When pomps shall in a long procession pass,
> Wreath'd on the posts before his palace wait,
> And be the sacred guardian of the gate.
> Secure from thunder, and unharmed by Jove,
> Unfading as th' immortal powers above;
> And as the locks of Phœbus are unshorn,
> So shall perpetual green thy boughs adorn."

The true DAPHNE order have Laurel-like leaves and tough

bark, which is so acrid as to raise a blister on the skin. It used to be sold for this purpose in England, and is still on the Continent.

The Lace Tree belongs to this tribe, the delicate inner bark being used for the purpose indicated by the name.

The Spurge Laurel (Dáphne laureóla), is an evergreen shrub of two or three feet high; its glossy evergreen leaves are situated in clusters at the end of the stems. The flowers are also in clusters, small, green, with a four-cleft calyx and eight stamens; they are rather fragrant. The shrub grows plentifully on the Clink-bank, near Richmond, where my specimen was gathered.

The Mezereon (D. mezéreum, *Plate XIV.*, *fig.* 12), is a scarcer native species, though so abundant in gardens. When first I visited Swaledale, I heard from the country people of a sweet-scented shrub found in the woods early in spring. We sought continually for this fabled plant, and, at last, in the centre of a group of Hazels, close upon the banks of a mountain rivulet, we discovered the glaucous leaves and scarlet berries of the Mezereon. The flowers are of a bright pink and very fragrant; they appear in April before the leaf-buds are open. Close by its retreat were trees where the thrush had reared a large family that very spring, so that I was forcibly reminded of the poet's song—

> "Thou hast thy wish; all love to see
> Thy simple bloom, Mezereon tree!
> The thrush its sweetest minstrelsy
> Is pouring forth to welcome thee;
> Thy store of sweets the early bee
> Hath sought with ready industry;
> And, prizing much thy beauty, we
> Are come to greet thee joyously."

Here is a plant, which, though humble in itself, is the representative of a tribe of considerable importance. The Little Bastard Toad-flax (Thésium linophýllum), with its humble cluster of whitish-green flowers, and prostrate stems, is a mere weed, as are the other members of its family in Europe; but in

New Holland, the East Indies, and the South Sea Islands, the tribe counts its members among shrubs and trees.

The Sandalwood tree of commerce is one of them, it gives the name to the tribe. My little friend grows on Warminster Down, and near an old Roman encampment, called Battlesbury, on an adjacent hill.

The BIRTHWORT order has the perianth, or calyx, tubular, with a wide mouth. The first family of the tribe is the Birthwort (Aristolóchia); we have only one native species, and it is very rare. The flower is yellow, and swollen on one side, and the leaves are heart-shaped. I have only seen drawings of the plant. Some of the American species have enormous flowers of most quaint forms. In India boys wear the blossoms of one of these plants upon their heads instead of caps.

The Virginian Snakeroot belongs to this family. It is useful in medicine, having somewhat the effect of Camphor. It is employed in India as an antidote to the poison of snakes.

A specimen of the Asarabacca (Ásarum europǽum), was given me by Mr. Ward; it has a short stem, two kidney-shaped leaves on long footstalks, and a purplish calyx cut into three segments. It has one style and many stamens.

The Pitcher-plants are closely allied to the Birthworts.

When wandering on the Yorkshire moors I often found the trailing stems and short Yew-like foliage of the Crowberry (Empétrum nigrum), and I left an earnest charge with the gamekeeper's good-tempered wife to send me the blossoms in the early spring. No blossoms came; but when the fruit was nearly ready, she sent me a boxful of the plants. She wrote expressing her regret at finding no blossom. She had watched the plants carefully, she said, from the first opening of spring, and had ascertained for a fact that the fruit came without any blossom having preceded it. The flowers are certainly very small, and as the male and female blossoms grow on different plants, and the latter are particularly inconspicuous, we cannot wonder at the good woman's mistake. Children and moorfowl eat the fruit, but it is not very wholesome.

The Spurge order comes next, terrible in its power for evil, though in some respects it is serviceable. It is a perplexing tribe, for in some members the petals seem perfect, whilst others have neither calyx nor corolla. The stamens and pistils grow on separate flowers. The members of this tribe vary from mere weeds to trees of the largest size. They have all an acrid juice, which in some is virulent enough to be a most dangerous poison; while in others the evil quality is dissipated by heat, and the farinaceous substance remaining becomes wholesome food for man: the Cassava is an example of this.

The Manchineel tree, whose very shade is dangerous, belongs to this tribe, and other members produce croton and castor oil, cascarilla bark and indiarubber; and the "vegetable tallow," of which candles are made in China, is also formed from a member of this order.

The British members of the Spurge family are green leafy plants, but many foreign species are leafless, with enlarged stems like the Cactus.

The Petty Spurge (Euphórbia péplus), even, the mere garden weed which children call "Fat-hen," can raise a blister with its milky juice.

The Caper Spurge (E. láthyrus), which we often see in gardens, where it is cultivated for its caper-like seeds, abounds in poisonous juice. The stem is tinted with crimson, and its leaves are glaucous. It is very dangerous to use the seeds instead of capers. Fée states, that with as much oil drawn from this plant as may be bought for a franc ninety persons might be poisoned.

The poor fishermen of Kerry stupefy the fish in the rivers by throwing a basketful of the Irish Spurge (E. hibérna), into the water. This plant has six forked branches, and its flowers are purple. We have no specimen of it.

The Sun Spurge (E. helioscópia), is often called Wartwort,

for its milky juice is an effective caustic for removing warts. It differs from the Petty Spurge in its darker leaves and five branches, but it is also a weed in cultivated ground.

The Cypress Spurge (E. cyperíssias), is a common border plant in old-fashioned gardens. Its foliage is very small and glaucous. This is one of the most poisonous of our British Spurges.

Edward has the Dwarf Spurge (E. exígua). It is a pretty little tidy plant, with narrow glaucous leaves, and flowers in three branches. It grows in corn fields about Hawkhurst.

The Portland Spurge (E. portlándica), has also narrow leaves, but it has five branches. I gathered it on the rocks by the Loe Pool in July.

The Wood Spurge (E. amygdalóides), covers acres of ground in the Herefordshire woods, especially where the trees have been much cut. At a little distance it looks like a broad stretch of sunlight. It grows freely between Plymouth and Looe on the roadside; also, in woods in Somersetshire and Kent. It is a shrubby plant, with six branches, and large, blunt, hairy leaves.

There are an Upright Warty Spurge (E. stricta), with dotted capsules and pale foliage; and a Purple Spurge, with procumbent stems, frequenting sandy seashores. Both of these are rare, and we have no specimens.

The Mercury family is nearly as poisonous as the Spurge.

The Dog Mercury (Mercuriális perénnis), is a dark green herb, with upright stems, closely set with lance-shaped serrated leaves. The stamens and stigmas are on different plants, and the green flowers are arranged in spikes.

The Annual Mercury (M. ánnua), has branched stems, and is of a lighter hue. My acquaintance with this was made in botanic gardens only.

The honest Box tree (Búxus sempervírens), is the only plant of this tribe that can be said to have a really good character. It grows frequently in shrubberies, and is said to be wild also.

The stigmas and stamens are in different flowers, but on the same plant. People dispute as to whether this is a British shrub, but at any rate it has been naturalised long enough to give names to places; there is a Boxley in Kent, and Boxhill in Surrey. Even in old Gerarde's time it was considered a British tree, for he says, "It groweth upon sundry waste and barren hills in Englande." Like the Yew, this tree used to be tortured and clipped into marvellous shapes, but that fashion is past. Box wood is the best material for wood-engraving, and for mathematical instruments. It is so heavy that it will sink in water. The sight of it brings the pleasant Yorkshire Christmas to my mind, when we have Box, Holly, and Yew in every window, and hung in garlands round every picture, and a row of red Apples stands on the mantlepiece, one for every member of the family, with a shrubby branch of Box stuck in the middle of it.

We now come to the Nettle tribe; and though at first sight we may consider that it will complete our large group of unattractive plants, yet I fancy we may find a greater amount of interest here than we expected. The tribe contains three British families—the Nettle, the Pellitory, and the Hop; and its foreign members include the deadly Upas, the delicious Fig, and the forest Banyan.

Of our three native species of Nettle the Roman Nettle (Úrtica pilulífera), has the most virulent sting. It bears its flowers in round heads, and grows chiefly in the neighbourhood of the sea.

The Small Nettle (U. úrens), frequents the environs of villages; indeed my specimen was gathered from the very centre of a hamlet in Swaledale. It has oval leaves and simple clusters, and its sting is less biting than that of its brother above mentioned.

The Great Nettle (U. dioíca), is the commonest and least offensive species. It grows everywhere, and its young shoots used to be eaten generally as a dinner vegetable; cooked as

Spinach it forms an agreeable and wholesome dinner dish. Any botanist rambling in the lanes in spring with a basket and gardening gloves may win the favour of his or her hostess by bringing back young Nettles for the pot. Sir Walter Scott makes Andrew Fairservice boast that at Dreepdaily, where he was brought up, they grew "Nettles for Spring Kail."

There are foreign Nettles as large as trees, some of which are without stings, while others are so virulent that their slightest touch occasions intense agony. The fibre of the Nettle is both fine and strong. Thread and lace are made of it in Java and Assam, and I have seen bunches of it in the museum at Kew which resemble glossy spun silk. Grass linen is formed from Nettle fibre.

Our one native species of Pellitory, the Wall Pellitory (Parietária officinális), grows on the ruins of Richmond Castle. It has a red stem and lance-shaped leaves, with pinkish flowers in their axils. It has generally a dusty appearance, for its hairs attract any loose grime that may be floating in the air.

The Hop (Húmulus lúpulus, *Plate XIV.*, *fig.* 13), is Edward's favourite plant. He is enthusiastically enamoured of the Hop gardens of Kent; he says the bine-like plants, climbing ten feet high to the top of their long poles, and then hanging in tendril-like stems downwards, or stretching to meet the extended arms of the next Hop plant, or *bine*, as they are called, form a picture of beauty impossible to excel in the vegetable kingdom. Add to this grace of foliage the not unfrequent accompaniment of an Irish group, a coarse basket containing the baby slung between two of the poles in a very bower of Hops, the father and mother diligently picking the fragrant cones from bines laid across a large measure, and three or four bare-legged children assisting in their work, with more or less assiduity, as their wild nature inclines them, and you have a picture which a painter might build his fame upon. The Hop plants are cut down in the autumn, and left bare all winter. In the spring and summer the shoots are carefully tied from time to time, and the ground weeded, and in due time the male and female flowers open on their separate plants. Then the cone-like clusters form, and the bines grow tall, and become more graceful every week. When the Hopping season has come, every man, woman, and child forsakes the villages, of course excepting those who have some pressing occupation. Invalids are dragged to the Hop gardens to breathe the salubrious aroma, and all hands apply to the work. A whole family pick into one measure, and each little child can earn something. The Hops are carried from hence to an oast-house, where they are spread upon a grating, underneath which stoves are burning. When sufficiently dry, they are closely packed in sacks, and are ready for the market. The bracts of the calyx are studded with glands containing a bitter aromatic principle, on which the virtues of the Hop depend, and which gives the flavour to beer. The plant is wild in many parts of England—Wiltshire, Warwickshire, Yorkshire, &c.

The Hemp, representing another family in this Nettle order, can hardly be called a native plant, though it is cultivated in some districts for its fibre, from which cordage, ropes, and coarse linen are manufactured. The seed is a favourite food for birds. The flowers have the stamens and pistils in different blooms, and the wand-like stems and palmate leaves look stately enough.

CHAPTER XV.

ULMÁCEÆ — SALICÁCEÆ — BETULÁCEÆ — MYRICÁCEÆ — CORYLÁCEÆ — CONIFÉRÆ — TAXÁCEÆ.

> " The Oak tree and the Mountain Pine,
>   The Willow on the fountain's brim,
> The Birch tree and the Eglantine,
>   In reverence bend to Him;
> The song-birds pour their sweetest lays
>   From tower, and tree, and middle air;
> The rushing river murmurs praise—
>   All nature worships there!"
>
> DAVID VEDDER.

FROM the time when Adam "heard the voice of the Lord among the trees of the garden" to the present day, these lords of the vegetable creation have borne a marked part in the history of man. When Israel was oppressed by enemies, and Gideon threshed his Wheat in secret, an angel of the Lord came and "sat under the Oak tree that was in Ophrah," promising aid and deliverance. Elijah, exhausted in body and mind in the great conflict with idolatry, slept a miraculous sleep, and was miraculously fed under a Juniper tree. Trees were used in building from the earliest period, and they entered largely into the formation of the Tabernacle and the Temple. And, as evil is ever striving to overgrow the good, so the Creator's gift of trees was perverted to the use of idolatry. " He heweth him down Cedars, and taketh the Cypress and the Oak." "He kindleth it and baketh bread; yea, he maketh a god and worshippeth it. He burneth part thereof in the fire; with part thereof he eateth flesh, and the residue thereof he maketh a god." Idol-altars were raised beneath the covert of the grove. " Inflaming yourselves with idols under every green tree." "And they set them up images upon every high hill, and under every green tree." The Prophets have used trees as an emblem of creatures blessed by God. "As trees of Lign-aloes which the Lord hath planted, and as Cedar trees beside the waters." "And he shall be like a tree planted by the rivers of water, that bringeth forth his fruit in due season." "And they shall spring up among the grass, as Willows by the watercourses." A tree was the cause of the curse, and upon a tree was that curse atoned. "As it is written, Cursed is every one who hangeth on a tree." In the restored earth trees are represented as taking a prominent part. "All the trees of the wood shall rejoice before the Lord when he cometh." "Break forth into singing, ye mountains; O forest, and every tree that is therein." "All the trees of the field shall clap their hands."

We have all felt the soothing influence of an hour in the woods; the quiet and seclusion, together with the companionable shelter of the trees, bring a sweet calm to the overwrought mind. America's genial poet well describes this quiet restful influence:—

> "The green trees whispered mild and low,
>   'Come, be a child once more!'
> And waved their long arms to and fro,
> And beckoned solemnly and slow:
> Oh! I could not choose but go
>   Into the woodlands hoar:
>
> "Into the blithe and breathing air,
>   Into the solemn wood,
> Solemn and silent everywhere!
> Nature with folded hands seemed there,
> Kneeling at her evening prayer!
>   Like one in prayer I stood."

The ELM order stands the first in this our tree group. They are noble trees, and though not very early in coming into leaf, yet present a cheerful appearance in spring, from the quantity of green membranous capsules which quickly follow the clusters of brownish-purple flowers. These flowers have five stamens and two styles.

The name of the common Elm (Úlmus campéstris), carries

the mind back to many a pleasant wood, or cottage overshadowed by that beautiful tree. It has rough leaves, doubly serrated, and deeply cloven capsules. The Romans used to train young Elms into convenient shapes, and plant the vines to twine around them.

The Wych Elm (U. montána), has larger leaves, and the seed-vessels are less cloven. This tree is frequently met with in Scotland.

There is a fine avenue of the common Elm (U. campéstris, *Plate XV., fig.* 1), in St. James's Park, London, and another in the Champs Elysées, in Paris. The tree is mature at the age of 150 years.

There are a Cork-barked Elm, and a Smooth-leaved Elm, and a Cornish Elm; but these trees are rarely to be met with, and we have no specimens of them.

The CATKIN order succeeds that of the Elm, and includes a great many British families. Of these the Willow is the first. The flowers are arranged in catkins, or, as the children call them, palms; the male catkins on one plant, and the female on another.

By the side of the Swale, near Richmond, a great variety of Willows are to be found. There are the Osier Willow (Sálix viminális), so useful in basketmaking; and the Rose Willow (S. hélix), of low growth, and with pinkish catkins and narrow leaves; and the Basket Willow (S. Forbiána), the female of the Bayton Willow, both occasionally used for weaving. Opposite to these grows the Two-coloured Willow (S. bícolor), the whitish lining of its broad dark leaves being considered as a second colour; whilst by the side of a well, in a field not a hundred yards further off, towers a handsome tree of the Silky-leaved Willow (S. Smithiána), its slender greyish catkins closely resembling the narrow young leaves.

In the woods bordering this river the golden catkins of the Round-leaved Sallow (S. cáprea, *Plate XV., fig.* 2), make a gay object in the spring; and the Water Sallow (S. aquática), with its downy leaves and bright catkins, strikes its roots into the stream.

The Dwarf Willow (S. répens), creeps upon the moors near Middleham, entangling its branches with those of the Heath and Crowberry; and the Long-leaved Triandrous Willow (S. triándra), adorns the riverbanks near Easby with its graceful branches, light shining foliage, and long, slender, primrose-coloured catkins.

The Rusty-branched Willow and the Rosemary-leaved Willow (S. Doniána and rosmarinifólia), were given to me by Mr. Ward. He, in conjunction with the Rev. John Leefe, has made a first-rate collection of Willows, which he has published for the instruction and pleasure of the botanical world.

Edward has the Varnished Willow (S. decípiens). It grows near "Cockshot gate," at Hawkhurst. Its bark is shiny, and its delicate leaves and graceful catkins resemble the Long-leaved Triandrous species.

A handsome tree of the Bedford Willow (S. Russelliána), grows by a small farm-house in the same neighbourhood, and he has found the Crack Willow there too; it is distinguishable by the extreme brittleness of its branches.

The common White Willow (S. álba), grows to a good-sized tree, with greyish foliage. It is also found about Hawkhurst. It is rapid in its growth, and produces a great bulk of timber in a short time. When well ventilated and dry the wood lasts well, and has been found in a sound state in buildings a century old.

The pretty little Eared Sallow (S. auríta), having ears by the side of its leafstalks, and roundish yellow catkins, grows frequently on plots of ground by the roadsides, there called "shores."

In some of the woods I have found the Prostrate Willow (S. prostráta), growing in an entangled mat.

The Willow family numbers sixty-four species, the smallest of all being only a couple of inches in height, but with all the accompaniments of a tree.

The Weeping Willow is a foreign species, a native of

Babylon. It is the species alluded to in the Psalm: "As for our harps, we hanged them upon the trees that are therein." It was introduced into England by the poet Pope, who planted one of the twigs in which a quantity of Figs had been packed, and forthwith it took root.

The wood of the Willow family makes the best charcoal, and is therefore used in gunpowder. An extract from the bark is called salicine, and rivals quinine as a tonic. Martin Tupper speaks of this: "Not long to charm away disease has the Willow yielded its bark." The wood is useful for handles for light implements, as hay-rakes, &c. Hoops for barrels are made by splitting Willow-rods in two equal parts, and crayons are made of the charcoal. The Willow is the emblem of deserted love. Spencer speaks of it as

"The Willow, worn by forlorn paramour."

The Poplar family numbers four British species.

The White Poplar (Pópulus álba, *Plate XV.*, *fig.* 3), has leaves both lobed and toothed. In youth they are white on both sides, but the upper surface becomes full green. It is an elegant tree, with horizontal down-curving branches, and large catkins, the crimson stamens of which are very conspicuous. Like the Willows, these trees have the male and female catkins on different plants. I believe this is the tree into which the weeping sisters of Phæton were transformed. I will repeat the passage; but beware of shedding tears over the recital, lest you also should turn into Poplar trees.

"Four times revolving, the full moon returned,
So long the mother and the daughters mourned;
When now the eldest, Phaëthusa, strove
To rest her weary limbs, but could not move;
Lampetia would have helped her, but she found
Herself withheld, and rooted to the ground;
A third in wild affliction, as she grieves,
Would rend her hair, but fills her hands with leaves;
One sees her thighs tran-formed; another views
Her arms shot out, and branching into boughs."

Both this tree and the Black Poplar grow about Hawkhurst.

The latter has triangular leaves, of a light green, and long lax catkins.

The Grey Poplar (P. canéscens), has roundish-waved leaves, and its branches are more upright than those of the White Poplar, which it otherwise resembles. I have no specimen of it.

The Aspen, or Trembling Poplar (P. trémula), grows near Frome in Somerset, and I have a specimen from thence. It is a lofty conical tree, and its leaves are always quivering. There is a beautiful, though of course fabulous, legend regarding this tree, which is well expressed in a poem signed "F. C. W.," published in "Chambers' Journal." The story relates that at our Saviour's death all the trees bowed their heads with one exception :—

> "But one tree was in the forest
>   That refused to bow;
> Then a sudden blast came o'er it,
>   And a whisper low
> Made the leaves and branches quiver—
>   Shook the guilty tree;
> And the voice was, 'Tremble ever
>   To eternity :
> Be a lesson from thee read—
>   He that boweth not his head,
> And obeyeth not his Maker, let him fear eternally!'
>
> "So thou standest ever shaking,
>   Ever quivering with fear,
> For the voice is still upon thee,
>   And the whisper near,
> Like the guilty, conscience-haunted.
>   And the name for thee
> Is 'the tree of many thoughts,'
>   Is 'the tree of many doubts;'
> And thy leaves are thoughts and doubtings, for thou
>   Art the sinner's tree."

The BIRCH order comes next.

Coleridge calls the common Birch "The Lady of the Woods." It bears its male and female flowers in different catkins, but on the same tree (Bétula álba, *Plate XV.*, *fig.* 10). Its light sprays and silvery bark are familiar to every one.

The Weeping Birch is a Scotch variety of this. It adorns the romantic heights and "passes" in the Highlands.

The Dwarf Birch, too, is a stranger to me. It has round leaves, and only attains the height of two feet.

> "In fairy glen of Woodilee.
>   One sunny Sunday morning,
> I plucked a little Birchen tree
>   The spongy moss adorning;
> And bearing it delighted home,
> I planted it in garden loam,
> Where, perfecting all duty,
> It flowered in tasselled beauty.
>
> "When delicate April in each dell
>   Was silently completing
> Her ministry in bud and bell,
>   To grace the summer's meeting,
> My Birchen tree of glossy rind,
> Determined not to be behind;
> So with a subtle power
> The buds began to flower.
>
> "And I could watch from out my house
>   The twigs with leaflets thicken,
> From glossy rind to twining boughs
>   The milky sap 'gan quicken.
> And when the fragrant form was green,
> No fairer tree was to be seen
> All Gartshore Woods adorning,
> Where doves are alway mourning."
>
> DAVID GRAY.

Birch wood is of no great use in Britain. Brooms are made of it, and hoops for barrels; also poles. It makes excellent charcoal, and is used in the north for the soles of clogs. In America it is much more important. Canoes are made of the bark, and the wood is employed for building purposes, and the soles of shoes are fastened on by pegs of it. A very important use of small Birch wood is in making the reels for sewing cotton, so familiar in the work-basket of every needlewoman, rich or poor. The Highland glens are being rapidly shorn of their beauty, because of the wood being cut for this purpose. The peculiar odour pervading Russian leather is imparted by an oil drawn from the Birch tanning. In Norway the bark is used for soleing shoes, and in Lapland overcoats are made of it. My specimens are from the Birch-woods of Swaledale, not very far from Rokeby, where Sir Walter Scott celebrated the praises of this and its companion trees :—

> "Hoary, yet haughty, frowns the Oak,
> Its boughs by weight of ages broke;
> And towers erect in sable spire
> The Pine tree, scathed by lightning fire;
> The drooping Ash and Birch between,
> Hang their fair tresses o'er the green;
> And all beneath at random grow,
> Each coppice dwarf of varied show."

The Dutch Myrtle, or Sweet Gale (Myrica gále), is an odoriferous shrub, with its male and female catkins on different plants. It grows three feet high, and flowers in May. The leaves are lance-shaped. I saw quantities of it about the shores of the Scotch lakes and upon the hillsides last autumn. The catkins were long passed, but the fragrance of the foliage atoned for the absence of blossoms.

Here is the Alder (Álnus glutinósus), the one British member of its family. This tree grows in watery places, and many of our humbler streams flow along an "Alder-curtained bed." You see the fertile catkins are round, and remain on the tree till they become quite woody; the barren catkins are slender, and soon fall off. The leaves are roundish, and rather sticky. The wood of this tree is soft, but when kept under water it is very durable, and is therefore often used in the foundations of bridges. Virgil says that the first boat was made of this wood :—

> "Then rivers first the hollowed Alder knew."

And Homer also vouchsafes a notice of it :—

> "Around it and above for ever green,
> The bushy Alders form'd a shady screen."

This Beech blossom Edward gathered at Elm Hill, near Hawkhurst (Fágus sylvática, *Plate XV.*, *fig.* 4). The leaves had just attained their full size, but were yet of a tender green, and fringed with down. The catkins were round: the male ones on long stalks, the female ones sessile. Beech wood was formerly very much used for goblets, platters, scabbards, &c. Cowley mentions it:—

> "Hence in the world's best years the humble shed
> Was happily and fully furnished.
> Beech made their chests, their beds, and their join'd stools;
> Beech made the boards, the platters, and the bowls."

Boys eat Beech-nuts when they cannot get Hazel-nuts, and they form good food for pigs.

A great many Chestnuts grow about Hawkhurst. The large leaves of the tree are lance-shaped, and very evenly toothed; they are of a bright green. The male and female flowers grow on the same spike, the latter at the lower end of it. The fruit is contained in a very prickly calyx, resembling a hedgehog. All kinds of good things are made out of Chestnuts. First of all, they are not bad eaten raw; secondly, they are excellent roasted; thirdly a splendid soup is made of them, ditto curry; and fourthly, you cannot cook them amiss. The Italians make porridge of them, and call it Polenta, and I should have no objection to sup with them upon it; they also use them as Potatoes. The Latin name, Castanea, is derived from Castanium, a town in Thessaly, where these trees grew abundantly. Chambers' poet thought very highly of the Chestnut. He writes:—

> "Thou, O Chestnut! richly branched,
> Standest in thy might;
> Rising like a leafy tower
> In the summer light;
> And thy branches are fruit-laden,
> Waving bold and free,
> And the beams upon thee shed
> Are like blessings on thy head:
> Thou art strong, and fair, and fruitful, for thou
> Art the good man's tree."

The wood of the Chestnut is very valuable, closely resembling Oak, and only second to the Oak for carpenters' use. It is grown in Kent for Hop-poles (F. castánea, *Plate XV.*, *fig.* 5).

Who does not love the Hazel? Little children rejoice over

its "lamb tails," as they call its male catkins. The female catkins are so small as to be generally overlooked; they consist of a little tuft of crimson stigmas, growing closely to the side of the two-year-old branches. Boys love nuts, don't they just? and squirrels are so fond of them that, shy though they are, they will come into gardens where nuts abound. My friends at Hawkhurst have a grove of Filberts, and a squirrel was seen in the high trees adjoining on several occasions last autumn. The nuthatch feeds on the same fruit, wedging the nut in a forked branch, and then splitting it with its beak. While the nuts are very young a small insect pierces the shell, and lays an egg there. As soon as the kernel is formed, the grub begins to feed on it, and when the ripe nut falls the little creature gnaws its way out. This is the cause of the empty shells we find under the Hazel bushes. The Scarlet-cup fungus grows on decaying branches of this tree. In many of the mining districts the Hazel is still believed to have an affinity for metals. The diviner holds a rod across his breast, one end slightly inclined to the earth; when metal lies beneath, the rod is forcibly drawn downwards. The many resultless borings in such districts speak to the value of the test (Córylus avellána, Plate XV., fig. 7).

The Hornbeam (Carpínus bétulus), is pretty frequent in the hedges about Hawkhurst. Its bark is smooth, and its leaves hairy, resembling those of the Elm, except in their tint, which is paler. The fruitful catkins are placed nearer to the end of the branches than the barren ones; they are interspersed with large bracts, which give the fruit the appearance of very elongated Hop-clusters. The wood of this tree is the toughest of all wood, but as it does not take a good polish it is not in great request.

Last in the Catkin tribe, but certainly not least, stands the Oak tree, the pride and strength of Britain. There are two native species; but the Sessile-fruited Oak (Quércus sessiflóra), with its downy leaves, is little noticed or valued in comparison with the essentially "British Oak." The Druids worshipped the Oak, maintaining perpetual fire of its wood. Once a-year the people had to extinguish all their fires, which were relighted from the burning wood on the sacred altar. This is the origin of our beloved Yorkshire custom of the Yule log; this revivifying brand was generally of Oak, but sometimes of Ash. The botanical name (Q. róbur, *Plate XV.*, *fig.* 6), means *beautiful tree.* Cowper delineates the life of an Oak in a few touching lines :—

> "Thou wert a bauble once—a cup and bell,
> Which babes might play with; and the thievish jay,
> Seeking her food, with ease might have purloined
> The auburn nut that held thee, swallowing down
> Thy yet close-folded latitude of boughs.
> Time was, when, settling on thy leaf, a fly
> Could shake thee to thy roots; and time has been
> When tempests could not.
> Time made thee what thou wert—king of the wood!
> And time hath made thee what thou art—a cave,
> For owls to roost in!"

But much as we ever venerate the old decayed Oak, with its gnarled branches and hollow trunk, it is to the trees which fall in their prime that our national love and gratitude are given.

> "Thou, O Oak! the strong ship-builder,
> For thy country's good
> Givest up thy noble life,
> Like a patriot in the strife,
> Givest up thy heart of timber, as he poureth out his blood."

Oak bark is used for tanning; it is taken in April. The Royal Oak is from an acorn of the old Boscobel Oak, where the then fugitive king used to conceal himself during the day, taking refuge in the house at night. The Parliament Oak is so called from King Edward I. having held a parliament under it in 1290; it is in Northamptonshire. In 1842 "the large Oak tree" was still standing, among whose branches the hero Wallace and his men concealed themselves from the English. The tree is at Ellerslie in Renfrewshire. "The King of the Woods" is a gigantic Oak near Jedburgh; it is said to mark the place where the border clans used to meet in old times.

Mr. Johns speaks of an Oak called the "Gospel Oak," at Stoneleigh; the title has doubtless the same origin as that of the "Gospel Field," near Ripon. On a certain day in the year the clergy and choir of the church used to walk the boundaries of the parish, stopping to read parts of the service at different points, and chanting as they went along; the field where they paused to read the Gospel is still called the "Gospel Field." The custom of dressing the houses with Oak on the 29th of May, in commemoration of the king's concealment in the tree and his restoration to the throne, is familiar to us all. Such associations, together with our reliance on our "wooden walls," and the great beauty of the noble tree, make the Oak a general favourite, and we warmly echo Tennyson's exclamation:—

> "Oh! flourish, hidden deep in Fern,
> Old Oak, I love thee well."

The insect tribes love it as much as we; one pierces the bark and makes what we call "Oak Apples;" another covers the under part of the leaf with green Currant-like galls; a third performs a similar service on the little flower-stalks; and a fourth ornaments the back of the leaf with Oak spangles, in each of which a young insect is concealed. The Oak galls of commerce, which form an important ingredient in ink and in dyes, are from a species of Oak indigenous in Asia Minor and the Levant. Cork is the bark of another species of Oak. The green caterpillar feeds on the leaves, the cockchafer and stag-beetle live among the branches, and the purple emperor butterfly considers the Oak his home.

The CONIFER order comes next, and is the last in the great Two-lobed class. It contains three British families—the Pines, the Yews, and the Junipers; the Cedars and Cypresses belong to this tribe. The stamens and pistils are in different flowers, and the fruit is generally a cone, though sometimes a berry.

Our one British member of the Pine family is the Scotch Fir (Pínus sylvéstris, *Plate XV.*, *fig.* 9). In England this tree is chiefly used as a nurse for other trees, sheltering them from the rude blasts; in Scotland it is a great and characteristic ornament of wild rocky heights. Sir Walter Scott fitly terms it the " Pride of the Highlands :"—

> " Row, vassals, row, for the Pride of the Highlands;
> Stretch to your oars for the evergreen Pine!"

This tree is useful for timber and firewood, and a coarse tar is drawn from it for common purposes.

The best tar is from the Carolina Pine. The Spruce Fir yields excellent pitch, and turpentine is drawn both from the resin of the Alpine Fir and from that of the Larch. Lamp-black is the soot of burnt tar. In Russia roads are made of the common Fir, and at Swedish funerals the way is strewed with Fir sprigs. Masts of ships are made of this tree in every country. The poor Laplanders are thankful to grind the cones into flour and make bread of it, or to mix the ground bark with oatmeal for pancakes. Beer is made from the twigs of the Spruce Fir, and the Romans used to flavour their wine with Fir cones. The inhabitants of the Oregon mountains use the seed of a kind of native Pine as food; the seed is large, sweet, and nutritious. The victors at the Isthmian games were crowned with Pine.

We do not need to go into Scotland to find the Yew in its glory (Táxus baccáta, *Plate XV.*, *fig.* 8). It grows freely in many of our Swaledale woods, and is particularly luxuriant about Studley and Fountains Abbey. The Seven Sisters Yews, close by the old abbey, are widely celebrated; they are supposed to have been full-grown trees in 1132, the date of the building of the abbey, and several of them are standing still. The ochre-coloured catkins of this tree are not attractive, but its crimson berries are very beautiful, contrasting with the deep green of the foliage. The hardness of the wood makes it suitable for turnery articles, but its chief utility was in the days of archery. English Yew bows won the victories of Cressy and Poictiers; and in Switzerland the Yew is called

"William Tell's tree," because his famous bow was made of it. Keats furnishes his Endymion with such a bow :—

> "Again I'll poll
> The fair-grown Yew tree for a chosen bow."

We frequently find this tree in churchyards, for it has always been accounted a memento of death. Shakspeare alludes to this :—

> "My shroud of white stick all with Yew."

The foliage of the Yew is accounted very unwholesome; animals eating it are affected as with poison. It is sometimes administered as a village medicine; the dried leaves produce but little effect, but a spoonful of the fresh twigs have been known to cause death.

The Juniper (Juníperus commúnis), is also a familiar tree on our hills; or, I should rather call it bush, for it is either prostrate, or but a very few feet high. Its catkins, or flower-clusters, resemble those of the Yew, and its berries are black. These berries are used in the distillation of gin, and a beer is made in France, called Genevrette, from them, mixed with barley. The leaves of the shrub are so stiff and pointed as to be almost prickly.

The Cedar wood of which pencils are made is the produce of an allied species of Juniper, commonly called the "Red Cedar." Frankincense is obtained from the Juniperus lycia; the Hebrews in olden times, and the Romanists in the present day use it in their incense. The Arabian frankincense is accounted the best.

Dr. Murray speaks much of the permanence of vegetation, and of trees in particular, as remaining for centuries as characteristics of the same district. He says: "The Palm, among the princes of the kingdom of vegetation, the Fig tree, and the Olive tree, still characterise the 'Land of Promise,' though Palmyra and Jericho are no more; the Cedar of Lebanon still vindicates its claim, by imprescriptible right, to the domains of its ancestry, though Tyre and her merchant princes are as though they had never been; Bashan is still celebrated for its Oaks, as in the times of the kings of Judah; the Olive still flourishes on its native soil, the acclivities of Olivet, as when it formed the chief retreat of a chosen band; the Amaryllis lutea, the 'Lily of the Field,' still decorates the fields of Palestine, as when the Prince of Life pointed to it as a pledge of His Providence; the Sycamore still springs up by the waysides, and skirts the shores of the Sea of Galilee, as in the reign of Herod Agrippa; and the Willow still weeps by Babel's stream, though Babylon 'is fallen, is fallen.'"

CHAPTER XVI.

MONOCOTYLEDONS — HYDROCHARIDÁCEÆ — ORCHID-ÁCEÆ — IRIDÁCEÆ — AMARYLLIDÁCEÆ — DIOSCORI-ÁCEÆ—LILIÁCEÆ—MELANTHÁCEÆ—JUNCÁCEÆ.

> " The Lily of the Vale, of flowers the queen,
>   Puts on the robe she neither sewed nor spun ;
>   The birds on ground, or on the branches green,
>   Hop to and fro, and glitter in the sun.
>
> " Now is the time for those who wisdom love,
>   Who love to walk in Nature's flowery road,
>   Along the lovely paths of spring to rove,
>   And follow Nature up to Nature's God."

WE are now to begin the second great class of plants, the *One-lobed class* (Monocotyledonous). Here the seed-vessel does not split, as in the members of the last class, but germinates from the end. There is a second permanent mark of distinction between the classes in the veining of the leaves, the veins being disposed in a *network* fashion in the Two-lobed class, and in a *parallel* one in the One-lobed class. A third point of distinction, and one even more valuable, because it can be examined when seeds and leaves are both absent, is in the stem. The trees of the Two-lobed class increase their wood by concentric layers ; each year a new layer is added, and the whole is enveloped in the bark. These are called *Exogens*, because they are *outside growers*. The bark consists of three parts—the outside skin, the solid part, and the inner bark. The two inner parts often increase with such rapidity that the outer cracks ; some trees throw off the bark from year to year. As soon as a tree-seedling begins to grow a few inches high, the centre of the stem becomes pith, and rays extend to the limits of the skin ; then a layer is formed round the pith, and called the sapwood, and layers are continually added to this during all the growth of the tree.

Trees of the One-lobed class increase from *within*, and are consequently called *Endogenous*. These have neither pith, rays, nor concentric circles, nor any bark, properly so called, only an outer skin. The bole of an endogenous tree consists of a number of woody cells, interspersed with bundles of fibres; and as the stem lengthens, the leaves on its summit send down more woody fibre, so that the tree becomes more closely filled round the margin of the trunk, and the heartwood remains light; on this account the wood of these trees is of little value as timber (*fig.* B). The outer skin is capable of

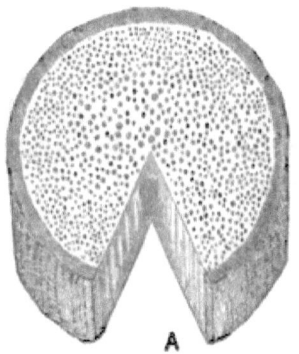 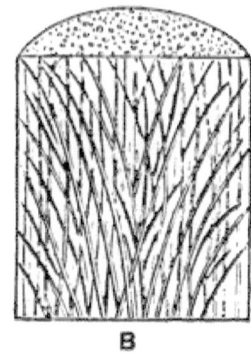

A Transverse section of the stem of an Endogenous tree.
B Longitudinal section of the stem of an Endogenous tree.

distention, so that the tree bole does increase in size to some extent, and the remains of the footstalks of the leaves help it to increase in height, while the new leaves of each spring send down fresh fibre to the trunk.

The third great class, the *Lobeless*, increase only by growing at the top; hence they are called *Acrogenous* or *summit-growers*. But we will not enter upon them at present. Our business of to-day lies with the One-lobed class, with their horizontally-veined leaves and inward-growing stems. Unfortunately our British plants of this class are almost entirely humble herbs.

The first order in the One-lobed class is the Petaliflors. Though characterised as thus, the petals are not always present, but the parts of the flower are arranged in a circular form, and the presence of petals seems to be provided for.

The first tribe in this order is the FROGBIT order, in which there are two families the Frogbits and the Water Soldiers. Both families have the stamens and pistils on different plants; three sepals, and three petals.

The common Frogbit (Hydrócharis morsus-rána, *Plate XVI.*, *fig.* 1), the one British member of the family, Fanny found in ditches about Kenmoor, near Clevedon. The delicate white flowers, which are very showy, grow three or four together, a transparent sheath enclosing the little stems at their junction with the main one. The leaves are pale, smooth, thick, and heart-shaped.

The only time I have seen the Water Soldier was in a pond near Goostrey, in Cheshire. We had planned going to that pond for some days, but there being a bull in the field where it was, we waited till we could get the escort of a gentleman. On reaching the pond our friend insisted on getting the plant for us, and while taking it from the water he cut his hand deeply with the sharp leaves; he said afterwards that he had no idea that botanising was so dangerous a pursuit. Dangerous is certainly not too strong a word to apply to the leaves of the Water Soldier; they are like knives for sharpness, and the points along the edges make them still more cruel. The flowers grow two or three in a cluster, like the Frogbit, and the petals are large and white; the flower-stems have a sheath where they join the principal one. The Water Soldier (Stratiótes aloídes, *Plate XVI., fig.* 2), grows in Duddingston Loch, near Edinburgh, and in Clunie Loch, Perthshire. There is a great difference between the two families in one respect, for the Frogbit has only three stamens, and the Water Soldier above twelve. The leaves of the latter plant are sword-shaped, and dark green; the flowers are half hidden among them.

The Valisnera family belongs to this group, but it has no British representative.

Two members of the family are interesting:—the Spiral Valisnéria, which has a coiled or spiral stalk, by which it can adjust itself to any depth of water, so that its flower may float on the surface, and afterwards contract, so as to draw down the seed-vessel to ripen under water; and the Sugar Valisneria, which is used to convey water to the sugar while undergoing the process of refinement.

I must here draw your attention to a new and extraordinary plant belonging to this family. The bargemen on the Faxton Canal have called it "Water Thyme," and botanists have gifted it with a name a yard long, Anácharis alsinástrum, (*Plate XVII., fig.* 1). I found it first during the midsummer holidays, when we were frequenting the neighbourhood of the Leam for the double purpose of boating and sheil-seeking. The water weed that we drew out in the hopes of finding molluscs upon it had whorls of leaves in three, with fringed edges. In vain I searched Withering for the plant, none answering the description could be found. It grows in such abundance in that sluggish stream that we frequently got our oars fast in it. For some time I cast about in vain for a history of my new friend, and at last I found all about it in a paper in "Chambers' Journal," entitled "Alarming Invasion." It appears that the plant was first discovered in a Berwickshire Loch in 1842, by Dr. Johnston; but its fame died away until Miss Kirby found it in the canal near Market Harborough in 1847. By this time the weed, growing with marvellous rapidity, had made its way out of Dunse Loch, where Dr. Johnston had found it, and was travelling down the Whiteadder to join the Tweed. In the same season a portion of it was discovered in a tributary of the Trent, and in 1849 it was "forming large submerged masses of a striking appearance" in the stream itself, and many of the canals communicating with it. A year afterwards it was discovered in the Cam, and in the course of a few months it so blocked up that river that extra horses had to be yoked to the barges to draw them through its masses. This plant, like the Frogbit, has its male and female flowers on separate plants, one only of which is found in Britain, though

T

both flourish freely in America. The minute flower has three sepals, tinged with pink and green, three flat petals, and from three to six stigmas. The footstalk is several inches in length, with a sheath at the base. The leaves grow in threes, egg-shaped, dark green, and sharply toothed, which causes them to cling. A large quantity of silex is contained in the cells along the margin of the leaf, which makes it so heavy that it is rather inclined to sink than to float in water. From the closely-clinging nature of the leaves they are often dragged along, and portions of the stem thus broken off. Each whorl has the power of rooting, and can grow as it travels down the stream, without requiring to adhere to the mud at the sides or bottom. It is a wicked and dangerous weed, and has been the death of more than one bather; and fishermen detest it, because it either drags away their lines altogether, or robs them both of hook and bait. No one has yet been able clearly to ascertain how the plant got into our British streams; it is possible that a seed or joint might have come in a crevice in the American timber which was employed in the Watford Docks. From hence the communication would be easy with the Oxford and Leicestershire canals, the Trent, and its tributaries. It may have spread into the Cam from the Botanical Gardens of Cambridge; and as many other water-plants are known to have been introduced into Dunse Loch, it is possible that this may belong to their number. At all events botanists have no hesitation in classing it among British flowers.

The ORCHIS order succeed the Frogbits, and are a much more numerous group. The flowers are arranged in a spike, and are all perfect. The calyx has three sepals; the corolla is divided into three parts—two petals, and a nectary or lip, the latter often having a spur behind. In hot countries most of this tribe are epiphytes, or parasites on other plants; but in temperate climates they grow, as with us, from the ground. They delight in warmth and heat, and are seldom found in dry cold places. They have generally only one anther, and instead of stigmas they have a sticky cavity. The fragrant Vanilla, which gives so delicate a flavour to sweetmeats, is the produce of an Orchis of that name; and the Salep of commerce, which is said to contain more nutriment than any other vegetable substance, is merely dried Orchis roots reduced to powder. Orchises are cultivated in the south of Europe for the production of Salep, and our common Orchis is as well fitted as any for the purpose. The beauty of the foreign Orchids, as exhibited in our hothouses, needs no comment. The Orchis family is the first in the tribe.

The common Purple Orchis (Órchis máscula, *Plate XVI.*, *fig.* 3), is familiar to us all, flowering early in May, and contrasting its rich spike of crimson-tinted blossoms with those of the Cowslip and Primrose. Its leaf, all spotted with purple, is very handsome. Its only drawback is its smell, which is very disagreeable.

The Green-winged Orchis (O. mório), greatly resembles the Purple one; the general hue of the blossoms is the same, but the two side sepals, instead of being coloured like the corolla, are striped with green. The flowers are fewer, and the offensive smell is quite absent. The leaves are of a light green, and not spotted. I have found it in abundance in the county of Durham, and Edward describes it as covering fields in Kent in June.

The Spotted Orchis (O. maculáta), grows everywhere in woods and field-borders. We have it from several counties. Its spike is very crowded, and the little pale flowers are prettily streaked and spotted with purple. Its leaves are spotted also, but the markings are much smaller than in the leaves of the Purple Orchis, and are more conspicuous, because of the paler tint of the leaf.

Kent also produces the Butterfly Orchis (O. bifólia). It has a very tall spike, and the flowers are placed at some distance from one another; this is a very good thing, for the lip is very long and narrow, and the spur also, and they need some space, or they would look quite confused. The colour of the flowers is creamy, and the scent is most delightful. I have gathered it in great abundance in hilly pastures both in Yorkshire and Cheshire.

The Pyramidal Orchis (O. pyramidális), I found in moist meadows in the county of Durham. The flowers are small, and arranged in a crowded cone: they have more of a rose tint than any other of our native Orchises. This flower has also an agreeable scent.

The Marsh Orchis (O. latifólia), is a Swaledale trophy. The spike is large and crowded, and the whole plant very strong. The roots of this species and of the Spotted Orchis are palmate, consisting of several elongated knobs. The Marsh Orchis has pale green, unspotted leaves.

The Aromatic Orchis (O. conópsea), grows freely in meadows about Thiernswood in Swaledale. Its spike is long, and less crowded than those of the two last-named species. It has no spots on its leaves, and its root also is palmate. It is exceedingly fragrant.

The White Orchis (O. álbida), has very small flowers, which are closely packed in a narrow spike. It is scentless.

The Frog Orchis (O. víridis), has fewer and more scattered flowers; they are larger, and of a brownish-green. Both these species grow in hilly pastures in Swaledale.

The Dwarf Orchis (O. ustuláta), I found on the Wiltshire downs, where it grows freely. I have also gathered it near Sutton in the Ripon neighbourhood. The flower is white, and the calyx blackened, which gives it a variegated appearance.

The Brown-winged Orchis grows in the chalky districts of Kent, and the curious Monkey and Lizard Orchises are occasionally found in similar districts, but are very rare. All these species have two knobs to their root; one of these produces the flower, and dies away afterwards; in the meanwhile, the other knob throws out an offset, which shall produce a flower in its turn, and another knob when the present one shall have performed its office in life, and decayed like its predecessors. This peculiarity makes the plant move a little every year, so that after several seasons are passed the place of the Orchis is at some distance from that which it first occupied.

The Man Orchis (Áceras anthropóphora), is a slender plant, with little insignificant flowers, and no spur. Its colour is greenish-white, with a yellow lip; it also frequents chalky ground.

The Musk Orchis (Hermínium monórchis), is a very small plant, with a spike of a few green flowers, without spurs, and which are very delicately scented. It favours chalky districts. Wiltshire produces these two curious Insect Orchises. The Fly Orchis (Óphrys muscifera, *Plate XVI., fig.* 4), looks like a stem with flies crawling up it, and the Bee Orchis (O. apifera), bears as close a similitude to a collection of bees.

The Spider and Drone Orchises are rare Kentish plants.

The generic name of this family of Insect Orchises is Ophrys, signifying to blacken; it is supposed that this is in allusion to a custom of the Roman ladies of blackening their eyebrows with these plants.

The Creeping Goodyera (Goódyera répens), is an Alpine plant, with white spiral spikes. We have not found it. But all the members of the Listera or Twayblade family are in my hands. Their calyx is spreading, nearly flat, and without a spur.

The common Twayblade (Lístera ováta), has two large coarse leaves resembling those of the Greater Plantain, from between which the slender spike of insignificant green flowers ascends. It is frequent in all our woods, and often in the fields adjoining them.

The Heart-leaved Twayblade (L. cordáta), is a very slender plant; its stem is a mere thread, and its tiny root seems to take no hold of the earth, only to rest lightly among the dead leaves and twigs. The small spike boasts eight or ten little brown flowers, and the pair of light green heart-shaped leaves, half way up the stem, form the principal part of the plant.

The Bird's-nest Listera (L. nidus-ávis), is a curious plant, almost like a Broom Rape. Its fleshy pale brown stem is clothed with scales, and the calyx and corolla are of a uniform brownish hue. There are no leaves, and the matted fibres of the root resemble a bird's nest. Both these species grow in the Swaledale woods; this one is more frequent than the Heart-leaved Twayblade, which I have only found in one place.

The Fragrant Lady's Tresses (Neóttia spirális), represents the next family. It grows on the downs near Brixton Deverill, in Wiltshire. The flowers are white, with green calyces and bracts, and no spurs; they all turn to one side, and the stem twists. The odour is very delightful.

The Large White Helleborine (Epipáctis grandiflóra), grows in woods in the Warminster neighbourhood. It is a tall plant, with a few large, white, half-open flowers. It is, I think, the prettiest member of the Helleborine family. The leaves are broad lance-shaped, and grow at intervals up the stem, which is a foot high.

Here is the Broad-leaved Helleborine (E. latifólia); it comes from a wood near Kenilworth. The spike is crowded with flowers—at first green, but becoming brown as they get older; they all turn to one side, and the ovate-pointed leaves clasp the stem. It flowers a month later than the White species, never opening its petals till August.

There is a Marsh Helleborine (E. palústris), with beautiful white flowers, which grows in a marsh near Skeeby, in the Richmond neighbourhood. It is a noble plant. There are a Narrow-leaved Helleborine, and a Purple Helleborine; the two first have white flowers, and the last purple ones. We are not fortunate enough to have specimens of any of them; they are all rare plants.

The same may be said of the two members of the Bog Orchis family, the last in the Orchis tribe; they have spreading petals and no spur. The Two-leaved species (Maláxis Lœsélii), is peculiar to Ireland and Jersey; the Least Bog Orchis to Ireland alone.

The IRIS order comes next, and includes, besides the family of its own name, those of the Íxia and the Crocus. The various hues in the flower have procured for the Iris its name, and it has many interesting associations. Our common garden Iris is the Fleur de Lis which Louis VII. adopted as his heraldic emblem when he set out on the Crusade; it is still the flower of France. I remember a rather ludicrous mistake, arising from a confusion of terms and significations, when we were visiting the Staubbach waterfall in Switzerland. The torrent falls over a beetling precipice of an immense height, and spreads into a sheet of spray ere it reaches the ground. Up to mid-day the rays of the sun fall upon this misty cloud, and paint it with rainbow tints. We were a little delayed upon the road, and an intelligent little girl of our party betrayed great impatience. I inquired the cause, and she replied, "Oh, if you don't get there soon we shall not see the great Water Lily! I have read in a book called 'Leila,' that there is a Lily by the Staubbach so large that she stood under it, and it can't be seen after twelve o'clock." "Are you sure the book did not say *Iris?*" I inquired. "I believe it did," was the reply; "but is not the Iris a Lily?" The poor child was sadly disappointed to find that her expected Lily was but a kind of a Rainbow. Imaginative people charge the poor Iris with the changeableness of its prototypes, and the poet Locke is one of these—

> " The ever-changeful Iris there
> I charge thee not to bring :"

The root of a Florentine Iris yields the perfume which we call Orris-root. In Britain we have only two wild species of Iris, both of which have three stamens and one stigma, and the corolla is in six petals.

The Yellow Iris or Water Flag (Íris pseudácorus, *Plate XVI.*, *fig.* 5), is very common in swampy places, and the margin of ponds and rivers. It is a stately plant, with sword-shaped, glaucous leaves, and well deserves its French name "La Flambe aquatique." Macgillivray states that the root is recommended as a cure for toothache, and that it is used for dyeing black in the Hebrides. Dr. George Johnstone says the berries are a good substitute for coffee.

I have the second species, politely called the Stinking Iris

(I. fœtidissima). It grows freely in Devonshire, frequenting thickets in many places. My specimen came from near Exeter. Its petals are pale and beautifully veined with purple. Its leaves are sword-shaped, and of a brighter green than those of its brother; and its seed-pod is very beautiful, showing, on bursting, a gay assembly of bright orange seeds.

Our one British member of the Ixia family (Íxia bulbo-códium), grows also in Devonshire. I have it from Dawlish Common. It somewhat resembles a minute Crocus, but the flower bends slightly downwards. There are five petals, three stigmas, and one style. The corolla is pale lilac, with a green tint towards the base. It is one of our rarest British plants.

The Crocus family is the last in this tribe which has British representatives. Some friends of mine who know the Ionian Islands describe the Crocuses as colouring acres of ground there with their golden and purple hue; but they are rare plants in England, and some called native species are very doubtfully indigenous.

The Golden Crocus (Crócus aúreus, *Plate XVI.*, *fig.* 6), is a familiar and welcome ornament to our gardens, opening its glorious blossoms as soon as the snow has departed. It has been found wild in Suffolk, but my specimens are garden ones. The poet interprets its feelings very pleasantly.

> " Down in the darkness under the snow,
>   Where nothing cheering can reach me,
>   Here without light to see how to grow
>   I'll trust to nature to teach me.
>
> " I will not be cross, or idle, or frown,
>   Locked up in so gloomy a dwelling;
>   My leaves shall shoot up, and my root shall shoot down,
>   While the bud in my bosom is swelling.
>
> " Soon from my heart shall gay petals diverge,
>   Like rays of the sun from their focus;
>   I from the darkness of earth shall emerge
>   A happy and beautiful Crocus.
>
> " Now, while you admire my leaves and my flower,
>   I pray you my counsel to follow,
>   If patient to-day, in your gloomiest hour,
>   You'll come out the brighter to-morrow."

The Saffron Crocus has a lilac blossom. The stamens form the saffron of commerce. The plant used to be cultivated to a great extent in the neighbourhood of Saffron Walden, thereby giving the name to the place. Women were employed to gather the flowers just before they were ready to expand; they then carried them home and took out the stamens. An acre of ground produced two pounds of dried saffron the first year, and twelve during the two second years; after that the bulbs required to be renewed. The saffron is no longer used in dyeing, as the colour so soon fades. It is still an ingredient in confectionary, especially on the Continent and in Turkey; and it is employed in medicine, but the demand for it being less, the article is very little, if at all, cultivated in England.

The Purple Spring Crocus (C. vérnus), resembles the Saffron, but instead of golden stigmas it has them of a pale lemon colour. It flowers in March. My specimen was sent from Norfolk.

Edward has the Naked-flowered Crocus from Warwickshire. There are meadows about Warwick that are quite gay with its lilac blossoms in October; and, late though it be, the leaves are still later in appearing than the flowers.

I believe the Gladioli of our gardens belong to this tribe, as do also the useful plants which produce the Ginger, common Arrowroot, and Tous-les-mois of commerce.

The AMARYLLIS order succeeds that of the Iris. In this order the flowers have three petals, and three coloured sepals; their stamens are generally six in number. Beautiful as are the flowers of this tribe, the roots of many of them contain dangerous poisons. In some species this poisonous quality can be dissipated by heat, and then the farinaceous substance which remains forms wholesome food for animals or man. The arrowroot of Chili is an example of this.

The first family in the tribe is the Narcissus family.

The common Daffodil, or Lent Lily, is by rights a Narcissus (Narcissus pseudo-narcíssus, *Plate XVI., fig. 7*). I have gathered this favourite spring flower growing wild in abundance near Little Ouseburn, in Yorkshire, in many parts of Herefordshire, and Edward describes it as half covering fields and groves in Kent. This surely was the manner in which those grew which Wordsworth describes:—

> "I wandered lonely as a cloud
> That floats on high o'er vales and hills,
> When all at once I saw a crowd—
> A host of golden Daffodils,
> Beside the lake, beneath the trees,
> Fluttering and dancing in the breeze.
>
> "The waves beside them danced, but they
> Outdid the sparkling waves in glee;
> A poet could not but be gay
> In such a jocund company;
> I gazed and gazed, but little thought
> What wealth to me the show had brought
>
> "For oft when on my couch I lie,
> In vacant or in pensive mood,
> They flash upon that inward eye
> Which is the bliss of solitude;
> And then my heart with pleasure fills,
> And dances with the Daffodils."

Herrick, too, loved the Daffodils, and likened their early fading to the shortness of human life:—

> "Fair Daffodils, we weep to see
>   You haste away so soon;
> As yet the early-rising sun
>   Has not attained his noon.
>   Stay, stay,
>   Until the hastening day
>   Has run
>   But to the even-song;
> And, having prayed together, we
>   Will go with you along.
>
> "We have short time to stay, like you;
>   We have as short a spring;
> As quick a growth to meet decay
>   As you, or anything.
>   We die,
>   As your hours do, and dry
>   Away,
>   Like to the summer's rain,
> Or as the pearls of morning dew,
>   Ne'er to be found again."

We used to love the Daffodils when we were little children, and sing to them on their first appearing:—

> "Daffydowndilly, just come from town,
> With a yellow petticoat and a green gown."

There are two other species of Narcissus; one, the Poet's Narcissus, has pure white petals, and a yellow nectary-cup in the centre, with a crimped red edge. The whiteness of the blossom is a puzzle to me, for in Ovid's story it is called yellow.

> "Narcissus on the grassy verdure lies;
> And, whilst within the crystal fount he tries
> To quench his heat, new heats arise;
> For, as his own bright image he surveyed,
> He fell in love with the fantastic shade."

So then, as the fable goes, he pines for the love of his own shadow, and at length dies, and Echo and the nymphs mourn for him.

> "And now the sister nymphs prepare his urn;
> When, looking for the corpse, they only found
> A rising stalk with yellow blossoms crown'd."

Other classic authorities declare the Narcissus of fable to be purple, so it is evident that, unless our colder climate has bleached the blossom, our "Poet's Narcissus" cannot be identical with that of Ovid. I have found the plant wild in Durham, near the margin of a pond, and one specimen near Ross. It is a welcome flower in our gardens.

The Two-flowered Narcissus is even more fragrant than the Poet's Narcissus. It is cream colour, with a yellow nectary. I found it wild in the Gallow Fields near Richmond.

The Snowdrop family is the next in order.

We have only one native member of it (Galánthus niválu). There is a copse near Ripon, called Skelcrooks, where Snowdrops used to grow in great abundance. We made several expeditions thither to get the bulbs to plant in our gardens, and it was a regular spring treat to go and see the Snowdrops. Westwood's idea of the mission of this lovely plant is very pretty:—

> "The Snowdrop is the herald of the flowers,
> Sent with its small white flag of truce to plead
> For its beleaguered brethren. Suppliantly
> It prays stern Winter to withdraw his troop
> Of wild and blustering storms; and having won
> A smile of promise from his pitying face,
> Returns to tell the issue of its errand
> To the expectant host."

Every one hails the Snowdrop with delight. The weary winter is passed, and its stern hand has bowed down many a beloved head whose hope of health is dependant on the coming of spring. A few of these pale blossoms carried to the sick bed, or placed beside the worn invalid, bring with them fresh hopes of new-sprung courage. Children rejoice in the white buds, and in their rapid sanguine nature feel that hosts of bright flowers will be open on the morrow, for that the spring has come indeed. There is a legend of the Snowdrop, a merely fanciful one, but very pretty. It describes the wonder and grief of Eve at first seeing a snow storm; an angel comes to comfort her, assures her that spring will return again, and catching some flakes of snow he turns them into flowers, as a pledge of the truth of his promise.

> "The angel's visit being ended,
> Up to Heaven he flew;
> But where he first descended,
> And where he bade the earth adieu,
> A ring of Snowdrops formed a posy
> Of pallid flowers, whose leaves, unrosy,
> Waved like a winged argosy,
> Whose climbing mast above the sea,
> Spread fluttering sail and streamer free."

There is still one family left for us to consider—the Snowflake family.

We have only one British species, the Summer Snowflake (Leucójum æstívum). It closely resembles the Snowdrop, but its stem is taller, and bears several flowers in a cluster, and the sword-shaped leaf is bright green, not glaucous. My specimen was gathered beside the Dever, in Wilts, but I am not certain of its being wild. It is indigenous in Surrey. The roots both of the Snowdrop and Lent Lily are poisonous.

The YAM order succeeds that of the Snowflake, and of this we have but one British representative. A foreign member of the order, from which indeed it takes its name, is the Yam, the root of which is edible, and resembles in flavour a sweet Potato. It forms a very important article of food both in the East and West Indies.

Our British member is the Black Bryony (Támus commúnis), a climbing plant, with heart-shaped, shining green leaves, and long clusters of pale green flowers. The blooms on the barren plant have six stamens. The fruitful plant is a beautiful object in autumn, with its large clusters of scarlet berries and twining stems wreathing the shrubs in the grove or hedgerow. It abounds in the neighbourhood of Hawkhurst, and is equally common in Yorkshire.

The Herb Paris, or True Love (Páris quadrifólia), flourishes in the Richmond and Swaledale woods. The four leaves placed in a whorl make a fanciful representation of a true-love knot: hence the name True Love. The plant has no leaves, except those in the whorl upon the stem, and these are generally four in number, but they vary to five, or even six, occasionally. The flower is green, with four long green petals, and eight stamens. The fruit is a large black berry, and is very unwholesome, not to say poisonous. This is the only British plant of the tribe.

The LILY order comes next, and numbers some of the most beautiful plants in the world among its members. Our British representatives are the least showy of the tribe, though many of them do not lack beauty. There are useful members also, British as well as foreign. The Aloe belongs to this tribe; it was used by the ancients both for embalming their dead and decorating their funerals. Nicodemus brought "a mixture of myrrh and aloes" to embalm the body of our Lord; and Solomon ranks the Aloe among spices: "spikenard and saffron, calamus and cinnamon, with all trees of frankincense; myrrh and Aloes, with all the chief spices." The intoxicating liquor called Pulque, in which the Mexicans delight, is drawn from an American species of Aloe. A valuable medicine is obtained from the Aloe. The Maritime Squill yields also an important drug. The Hemp plants of Africa, America, and New Zealand also belong to this order. The two former are made into articles of clothing, and of the latter bridges are formed, and ropes, &c. The Leeks of Egypt, after which the carnal Israelites longed, are members of the Lily tribe; also the Eschallot of Asia Minor, the Garlic of Sicily, and the Onion of all countries.

The first family in this tribe is the Aspáragus (Aspáragus officinális). Its chief British home is Asparagus Island, opposite Kynance Cove, Cornwall; but Fanny could not go there, so we have no specimen. The family is easily recognised by the narrow leaves, which fall off as soon as they begin to fade, and by the scarlet berries. Our edible Asparagus is a native of the Crimea, and is also wild in Siberia and in Japan.

The Butcher's Broom family is the second in the tribe. Our one representative is the nearest approach to a tree which we have in our British One-lobed class (Rúscus aculeátus). It is a curious little shrub, with very stiff sharp-pointed leaves, which butchers used to stick about their meat, in the expectation that the flies would spear themselves upon them. The flowers grow from the centre of the leaf, and are followed by sessile scarlet berries. The male and female flowers are on different plants. My specimens are from a garden, for I have never found the Butcher's Broom wild.

Yorkshire has the honour of contributing the Lily of the Valley (Convallária majális, *Plate XVI.*, *fig.* 8), the favourite member of the family to which it gives its name. I have wild specimens from woods near Thornton, and from Studley, in the Ripon neighbourhood; but never did I see Lilies of the Valley in such profusion as in the woods about Weinheim in the Duchy of Baden. The plant grows very freely in Wilts, in a part of the ancient forest of Selwood, now called Sowley Wood, and the friendly people in the neighbourhood make parties every year to go and gather them. These excursions they call Lily Pies. I accompanied one of these rambles; there were plenty of Lilies, and their blossoms seemed more graceful, and their fragrance more charming than in our garden specimens. The woods about Dunkeld and Blairgowrie produce abundance of this favourite flower. I agree wholly in the opinion of the Scotch poet—

> "Sweet flower o' the valley wi' blossoms o' sna',
> And green leaves that turn the cauld blast frae their stems
> Bright emblem of goodness, thy beauties I lo'e
> Aboon the king's coronet circled wi' gems.
> There's nae tinsel about thee to mak' thee mair bright;
> Sweet Lily, thy loveliness a' is thine ain,
> And thy bonnie bells dangling sae pure and sae light,
> Proclaim thee the fairest o' Flora's bright train."

The seed of the Lily of the Valley is a scarlet berry, and the spikes of these almost vie with the flowers in beauty.

In Norridge Wood, in the same district, I found the waxen bells, tipped with green, of the common Solomon's Seal. The plant grows two feet high, its broad leaves clasp the stem, and are ranged alternately the whole length of it; two or three drooping flowers are placed at each axil. The whole plant has a drooping form. It grows, also, in the woods about Looe, in Cornwall. There is a species with angular stalks and large solitary flowers, and another with narrow whorled leaves; but I have found no specimens of either.

The Squill family is the next in succession; they are pretty plants, with blue or lilac blossoms arranged in a spike or cluster.

Beautiful specimens of the Vernal Squill (Scilla vérna), were sent to me from the Isle of Man, and it abounds on the coast of Cornwall. The flowers are bright blue, in a full cluster, and the numerous thread-shaped leaves rise all around the stem.

The Autumnal Squill (S. autumnális), has a more lilac hue, and it is a stronger plant. These specimens are from Torquay;

it is to be found all along the coast of Devon and Cornwall It expands its blossoms early in the autumn, and the leaves spring when the flowers have faded.

The Two-leaved Squill (S. bifólia, *fig.* 4), is the prettiest of the set; it has a few bright flowers arranged in a spike, and the turn of the stem is very graceful. Its two leaves are broader than those of the other species, and often tinged with crimson. My specimens are from a garden.

The Wild Hyacinth (Hyacínthus nonscriptus, *Plate XVI., fig.* 9), abounds in woods in spring. I have never visited a locality where it was not to be found. People will call it "Bluebell;" and certainly its flower is bell-shaped and blue; but the true "Bluebell" is the Harebell. The bulb of the Wild Hyacinth grows very deep into the ground; its flowers have six petals and six stamens; and the whole plant abounds in a thick juice which was formerly expressed to make starch. The roots are unwholesome, and when used in mistake for Onions have caused serious illness. All my Yorkshire specimens are blue; but I have them of pink and white from that delightful remnant of Selwood Forest. The Latin name means *not written*. There is an old legend that " Ai (alas)" was decipherable on the Hyacinth leaves, but this species lacks the markings which in others were ingeniously construed into letters, and is, therefore, the "Unwritten species." I fancy that Percy Bysse Shelley must have been at a "Lily Pie," and gathered the Hyacinth of three colours, for he writes—

> " And the Hyacinth purple, and white, and blue,
> Which flung from its bells a sweet peal anew
> Of music so delicate, soft, and intense,
> It was felt like an odour within the sense."

It must have been of the pink variety that the story is told, or more probably of a deeply-tinted foreign species; it states that Apollo, " a being fraught with all earth's richest gifts," was bound in a close friendship with young Hyacinthus. Zephyr loved Hyacinthus too; but Hyacinthus had no room in his heart for any one but Apollo. Zephyr became mad with jealousy and hate, and murdered Hyacinthus. Then Apollo—

> "Made this scarlet flower
> Spring from the dismal flood, and on its leaves
> Impressed the words of grief, Ai, Ai!"

There is a Starch Hyacinth (Muscári racemósus), which yields more starch than the wild Hyacinth. The spike of flowers resembles a miniature cluster of Grapes; they have a scent like starch. It grows in several places about Edinburgh, and also in a valley near Durham; but as it is generally found where old gardens have stood, there is much doubt of its being a wild flower.

The Star of Bethlehem family comes next. It has four members—three with white, and one with yellow flowers; six petals, more or less tinted with green on the outside, six stamens, and no calyx.

The common Star of Bethlehem (Ornithógalum umbellátum), is a familiar plant in gardens, its pure white clusters of stars appearing before the leaves. I found it growing freely in fields near Goostrey, in Cheshire, flowering early in May. The blossoms close towards evening.

> "Pale as a pensive cloistered nun,
> The Bethlehem Star her face unveils
> When o'er the mountain peers the sun,
> But hides it from the vesper gales."

The Drooping Star of Bethlehem (O. nútans), is a beautiful plant; the large half-closed flowers hang from either side the tall stem. The buds are greatly tinged with green, and little stripes and stains of it remain upon the mature flower, adding to the beauty of the intense white of the petals. I first saw the plant in the botanical gardens at Edinburgh, and was eager to make interest with the officials for a specimen; but since I have found it wild near Fawcett, in the North Riding of Yorkshire.

My specimen of the Tall Star of Bethlehem (O. pyrenáicum, *Plate XVI., fig.* 11), comes from Freshford, in the Bath neighbourhood. The young buds were formerly sold as pot-herbs in the Bath market; it is very abundant in that district. The plants sent to me were upwards of a foot high, and the flowers were thickly set upon the stem for half its length. The petals have a cream-coloured hue, and are tipped with green; they are narrower than in the other species, and the flowers smaller. The general expression of the spike reminds one of a Butterfly Orchis. It blooms in June and July.

The Yellow Star of Bethlehem (O. lúteum), grows in fields off Wicliffe Lane, near Ripon. It is a smaller plant than any of its brethren, and has a pair of leaves set on with its flower-cluster as well as from the root. But few flowers grow in each cluster; they are yellow within and green on the outside. This species flowers earlier than its companions. We used to gather it when seeking for Sweet Violets in April.

Of the Garlic family we have very few species.

The Great Round-headed Garlic (Állium ampeloprásum), grows on the Steep Holmes. The Sand and Mountain Garlics (A. arenarium and A. carinatum), I have never seen; these three species have flat leaves.

The Streaked Field Garlic, and the Crow Garlic (A. oleráceum and A. vineále), have roundish leaves; the latter I have found both at Clevedon and near Warminster. It has a round head often bearing bulbs as well as flowers; and, in common with all those species I have named, it has one or two long leaves upon its tall stem (A. vineále, *Plate XVI., fig.* 10).

The Broad-leaved Garlic (A. ursínum), is by far the prettiest member of the family. Its flowers are white, very much resembling a Star of Bethlehem, and they grow in an umbel. I have seen people, unaccustomed to wild flowers, spring upon it with great delight and gather a handful of its blooms, praising their beauty all the while; but then comes the recoil—the Garlic scent is overpowering, and everything that touches the plant is strongly perfumed by it. A friend of mine who, though delicate in health, is perfectly free from any affectation, and is an eager botanist, had a piece of this Broad-leaved Garlic brought to her in a glass of water to paint. Before she had been engaged many minutes with her sketch she became so sick that it was necessary to remove the plant, or she would have fainted. It is very common in woods. This and the Chive Garlic have both naked stalks; the stem is triangular or shaped like an awl.

The Chive Garlic (A. schœnoprásum), grows on the coast of Cornwall, but Fanny did not find it. It has a round stem, and a cluster of purple flowers. It is often cultivated for the sake of its leaves, which make an agreeable pot-herb.

This specimen of Fritillary comes from Oxford, where it grows freely in meadows. It has a stem about a foot high, with two or three grass-shaped glaucous leaves growing upon it, and at the end a large drooping bell formed of six petals all dappled with chocolate, and white, and green. Old Gerarde calls it "the Chequered Daffodil," and describes it as having "narrow grassy leaves, among which there riseth up a stalk three hands high, having at the top a flower of six leaves, chequered most strangely, surpassing the most curious painting that art can set down" (Fritillaria meleágris).

The Tulip is the next family in the Amaryllis order, and it is a scarce plant. It is found sparingly in the vicinity of Knaresborough, and I have had specimens from Suffolk, and near Marlborough, in Wilts. It is a gorgeous flower of six large bright yellow petals. Like our garden Tulips, it is bell-shaped at first, but opens wide like a star before falling. The broad glaucous leaves resemble those of the garden species.

We now pass on to the Meadow Saffron Tribe, which consists in two families, each having but one member.

The Meadow Saffron (Cólchicum autumnále, *Plate XVI.*), is a very pretty plant, adorning the pastures in the autumn with its multitude of Crocus-like flowers. It differs from the Crocus in having six stamens instead of three; it has also six lilac petals and no calyx. The seed remains concealed underground during the winter, and in the spring the capsules appear, along with the large bright green leaves. These capsules are gathered to be sold to the druggist; and from them a powerful medicine is prepared, which is very useful in cases of gout and rheumatic affections. The Meadow Saffron grows abundantly in fields between Ripon and Aldfield, and in similar pastures in Swaledale; and Edward reports an equal abundance around Warwick. Cattle avoid this plant, and the French evidently suppose that it poisons dogs, as they call it "Mort aux chiens."

The Scottish Asphodel (Tofiéldia palústris), is the one representative of its family; it grows four or five inches high, and its small yellowish flowers are arranged in an oval head. We have no specimens of it.

The Rush order, with its three families of Rushes, Wood Rushes, and Asphodel, comes next. Few account the Rushes an interesting family, but we have a good opinion of them on our Yorkshire moors. When land is forming on boggy ground the Rush is a great assistance. Wherever it gets hold the earth becomes consolidated round it, and when you get into a swamp, and do not know where to step, fearing that the inviting moss will sink under your feet, you have only to descry a cluster of Rushes, and you know that there is a safe footing; or, as the country people say, "there is a bottom."

The common Rush (Júncus conglomerátus), with its dense head of flowers diverging from the stalk, is familiar to all; and the Soft Rush (J. effúsus), with its branched panicle, is scarcely less frequent. I have found these on wet ground wherever I have been.

Of the Rushes with leafy stems I have three species.

The Moss Rush (J. squarrósus), I found at Brimham Rocks. It has no leaf till very near the base of the stem, and its branched cluster of flowers have bright yellow stamens. It has numerous channelled leaves growing in tufts from the root.

The Sharp-flowered Jointed Rush (J. acutiflórus), has joints in the stem, and grows in woody places. It is a larger plant than the last.

The Shining-fruited Jointed Rush (J. lampocárpus), is distinguished by its large polished capsules. I found it near Little Ouseburn.

The Toad Rush (J. bufónius), is a branched plant, with slender stems, and leaves at the branches. The petals, or rather sepals—for the Rushes have a calyx but no corolla—are of a white membrane, with little green veins up them. In the full sunshine flowers expand, and then they look like tiny stars. The ditches by the roadsides in Kent abound with it.

Often the little Round-fruited Rush (J. compréssus), grows along with the Toad Rush. Its stars are brown, and its stem and leaves stiff. Both these species have leafy stalks.

The Mud Rush (J. cœnósus), Fanny found at Clevedon. It is about four inches high, with a branched cluster of flowers, and crimson stamens.

The Bog Rush (J. uliginósus), is a very delicate little plant, the smallest in the family. The stem is leafy, and bulbous at the base. The flowers are in heads, about three together; each plant has a very few heads. Fanny found it near the Loe Pool, in Cornwall.

The Hard Rush (J. gláucus), is slender, tough, and glaucous. Its panicle is smaller than that of the Soft Rush, and grows lower down in the naked stalk.

I believe I was wrong when I said the Bog Rush was the least species. It is the smallest with a leafy stem, but the Alpine Rush is the least of all.

There are also a Great Sea Rush and a Lesser Sea Rush (J. acutus and J. maritimus), with naked stems; but I have no specimens of them.

Among the leafy-stemmed Rushes there are the Three-leaved Rush, frequenting Alpine bogs; the Slender-spreading Rush, a native of Scotch mountains; the Dense-headed Rush, an inhabitant of Jersey; and the Blunt-flowered Rush, living in marshes. We have no specimens of any of these.

There are six Wood Rushes, and four of them are common.

The Field Wood Rush (Lúzula campéstris), opens its brown stars and displays its yellow anthers in our hilly pastures early in the spring. All the family have broadish grassy leaves, with long down upon them.

The Broad-leaved Hairy Wood Rush (L. pilósa), is rather a larger plant, and lighter in form. It adorns the woods whilst its brother is decorating the fields, and is equally common with us.

The Narrow-leaved species (L. Fórsteri), is very slender, much more scarce, but favours our woods occasionally.

The Great Wood Rush (L. sylvática), is very common with us. Its matted clusters of pale green leaves form thick plats, and remain like a deep covering of straw when the flowers are quite dead. The stem grows a foot high, with a handsome panicle of starry brown flowers. The whole plant is downy.

There are a Spiked Wood Rush, and a Curved Mountain Wood Rush (L. spicáta and L. arcuáta), but we have not found them.

The Bog Asphodel family has only one British member, and it is termed, not very suitably, the Lancashire Bog Asphodel (Narthécium ossífragum, *Plate XVI., fig. 12*). I first saw this plant at Rudd Heath, in Cheshire, and I have since found it among Ling on the Swaledale hills in great abundance; so that, though it may be a native of Lancashire, I know it to be a native of at least two other counties, and the abundance of its spikes of rose-coloured seed-vessels which I saw spangling the wet ground in the Highlands shows it to belong to Scotland too. The flowers have six petals of a golden yellow colour, and the stamens are beautifully fringed, with orange anthers upon their summits. The blossoms are arranged in a spike two inches long, the whole stem measuring five or six inches. The leaves are sword-shaped, and wrapping over each other, and there is a bract in the middle of the stem, and another near its base. The root is creeping and fibrous, and the general hue of the plant a glaucous green. It is a very pretty plant, and has an excellent effect when contrasted with the dark purple Ling and white tufts of Cotton Grass. It is interesting, too, as belonging to the family of the Asphodel, so famous among the ancients, branches of which graced their funerals and were planted over their graves. It still abounds in the plains of Apulea. Homer and many other of the classic writers speak of it. It is a native of Sicily, and was supposed to abound on the further side of Acheron, watered by Lethe's stream. Pope introduces it in his Ode on St. Cecilia's day :—

> "By the streams that ever flow,
> By the fragrant winds that blow
>   O'er the Elysian flowers;
> By those happy souls who dwell
> In yellow meads of Asphodel,
>   Or amaranthine bowers."

CHAPTER XVII.

BUTOMÁCEÆ — ALISMÁCEÆ — JUNCAGINÁCEÆ — TYPHÁCEÆ — ARÁCEÆ — ACORÁCEÆ — LEMNÁCEÆ — NAIDÁCEÆ — CYPERÁCEÆ.

> " And nearer to the river's trembling edge
> There grew broad Flag flowers, purple prankt with white,
> And starry river-buds among the Sedge;
> And floating Water Lilies, broad and bright,
> Which lit the Oak that overhung the hedge
> With moonlight beams of their own watery light;
> And Bullrushes, and Reeds of such deep green
> As soothed the dazzled eye with sober sheen."
>
> PERCY BYSSHE SHELLEY.

WE have now but a few orders of plants before we come to the Glumaceous order, of which the Grasses are the prevailing members. Of these intervening families the most part are water plants. The first order is that of the FLOWERING RUSH; it has but one British representative. This, and the order succeeding it, are distinguished by having both calyx and corolla.

The Flowering Rush (Bútomus umbellátus, *Plate XVII.*, *fig.* 2), is a very stately plant, growing four or five feet high, with long triangular leaves. The flowers grow in a terminal umbel, at least twelve in each cluster; they have three crimson sepals, three rose-coloured petals, nine stamens, and six styles. It is the only British plant with nine stamens. The flowers are as large as those of the Marsh Marigold. Old writers speak of it as "the Grassie Rush." It grows freely in the Avon between Warwick and Guy's Cliff, and it is an ornament of the ponds at Longleat, the seat of the Marquis of Bath.

The WATER PLANTAIN order is the second gifted both with calyx and corolla; it contains two families—the Water Plantain and the Arrowhead. The former have three sepals and three petals, six stamens, and numerous styles.

The stem of the common species (Alísma plantágo), rises two or three feet high, and then flower-stalks diverge in branched whorls. The blossoms are pale lilac, and the plant would be very handsome but that the petals fall so quickly, that never more than a third part of the flowers are open at once. The leaves are large, oval-shaped, and of a very delicate green. It grows in many of the ponds about Hawkhurst, flowering in July and August.

There are a Floating Water Plantain (A. nátans), with simple flower-stalks, and a Lesser Water Plantain (A. ranunculoides), with larger flowers and lance-shaped leaves, but they are only found on mountain lakes; and there is the Starry-headed Water Plantain (A. damasonium), which has clusters of small white flowers, heart-shaped leaves, and two seeds instead of one in each carpel. I have no specimens of these three kinds.

Both the Leam and the Avon boast the Arrowhead (Sagittária sagittifólia, *Plate XVII.*, *fig.* 3), among their charms. It is a beautiful plant, the only one of its family. The stamens and pistils are in different flowers, and the leaves grow from the root, and are arrow-shaped. The blossoms are arranged in whorled spikes, and each one is as large as a Primrose. The three petals are of a brilliant white, with a blotch of deep violet at its base. The root of this plant was once believed to be a cure for hydrophobia, but that opinion is exploded. Some species are cultivated in China for their roots, which are there accounted a pleasant vegetable.

The ARROW-GRASS order have three petals and three sepals; they have six stamens and three stigmas. The leaves are thread-shaped and channelled, and the tall spikes are narrow, beset with the small green flowers.

The Marsh Arrow-grass (Triglóchin palústre), grows in swampy fields near Richmond, in Yorkskire.

The Sea Arrow-grass (T. marítima), flourishes on the muddy

edges of the Looe River. It is distinguished from the Marsh Arrow-grass by having six styles instead of three, broader fruit, and more fleshy leaves.

The REED-MACE order succeeds the Arrow-grasses, and they all like to grow in several feet of water.

The Great Reed-mace (Týpha latifólia, *Plate XVII., fig.* 4), is a regal-looking plant, seven or eight feet high, an immense club of brownish flowers crowning the summit. It grows along with the common Reed in a pond at Hill Deverill, in Wiltshire, and the tall stems look like a regiment of soldiers. The flowers containing stamens are situated in the upper part of the spike, and those containing styles in the lower part. If a spike of this Reed-mace in full bloom be brushed near a lighted candle a flash of light ensues. This phenomenon is referred by some persons to an electric property in the plant, but I believe that it is only that the scattered pollen ignites. On the Continent the down of the spikes is sometimes used for stuffing pillows, and the root is occasionally eaten as salad.

There are a Lesser Reed-mace (T. angustifólia), of a slighter form, and with the leaves semi-cylindrical and channelled above; and a Dwarf Reed-mace (T. mínor), with very narrow leaves, and a space in the spike between the male and female flowers. Both these species are rare, and Sir J. E. Smith does not consider the last-mentioned a British plant.

The Burr-reeds come next.

The Branched Burr-reed (Spargánium ramósum), is a handsome plant, and grows along with the Arrowhead and Flowering Rush in the Leam and Avon. I have seen ditches full of it in Durham; but it was pointed out to Edward in Kent as a rare plant. The leaves are triangular at the base, and the flower-stalk is branched. The flowers in all the species are arranged in heads, the upper heads with stamens, the lower with styles. It is very pretty when in full bloom, for the stamens and styles are long, and the pollen yellow, so that the appearance is as if yellow feathery balls were scattered over the plant.

The Unbranched Burr-reed (S. símplex), I got out of a pond in Cheshire; its "balls" are larger and still more feathery, but as the stem is simple there are only a few clusters of flowers at intervals up its spike. Here the smallest cluster is at the apex.

The Floating Burr-reed (S. nátans), is a much smaller plant. Its fellows are often three feet high, but this only rises a few inches out of the water; and the whole length of its stem rarely exceeds a foot, though that of course depends upon the depth of the water in which it grows. The hue of all these Burr-reeds is very pale, but this is the palest of all. The stalk is simple, and contains generally only one barren head, and about two fertile ones. My specimen is from a pond on Rudd Heath, Cheshire.

The next family is an exception to the group of water plants, at least our British member is. In other countries Arums inhabit moist places, and the margin of rivers. The beautiful white Arum of our greenhouses is the "Lily of the Nile," and upon the shores of the Mediterranean it is so plentiful as to be used in decorating halls and assembly-rooms for festive occasions.

Our native species called by the various names of "Cuckoo Pint," "Lords and Ladies," "Parsons in the Pulpit," is a handsome and curious plant (Árum maculátum, *Plate XVII.*, *fig.* 5). It resembles the Reed-mace in the arrangement of its flowers, the male being on the upper part of the spike, and the female in a broad ring lower down. A large sheath enwraps the spike before it comes to maturity, and continues to shield it like a canopy until the flowers are over; then berries form, the sheath withers and disappears, and a spike of scarlet berries takes the place of the canopied flower-cluster. The leaves are of a full green, arrow-shaped, and very glossy; they are often spotted with purple. They are decidedly poisonous. Orfila mentions three children who ate of them, two of whom died after lingering sufferings, and the third was with difficulty saved. In Queen Elizabeth's time starch was made from the root; and by grating it and steeping it in water the poisonous quality was removed, and the flower was used as an article of food in Ireland during the time of the famine. Mr. Johns mentions an extraordinary and terrible species of Arum growing in the West Indies, called the Dumb Cane. When chewed it causes the tongue to swell, and destroys the power of speech. Its effects last several days.

The Sweet Sedge tribe (Orontiaceæ), has only one British representative, and here it is (Ácorus cálamus, *Plate XVII.*, *fig.* 6). The flowers are all perfect, and arranged in a spike. The stem becomes leafy at the foot of the spike, and rises in a flag-like bract over it, or droops backward. It is very aromatic. This is the "Rush" wherewith they used to strew the floor of Norwich Cathedral on festival days, and garlands were made of it and hung in various churches and over graves in the olden time. Major Calder Campbell has noticed these customs in some touching lines:—

> " O riverside!
> Where soft green Rushes bear dark flowers,
> And reedy Grasses weave dark bowers,
>   Through which fleet minnows glide;
> O riverbanks! let me from you convey
> Something to scatter in yon ancient minster grey!
>
> "O minster grey!
> Where graves of friends beloved are found,
> I come to thee with strewments; round
>   Each blade of Grass, each spray
> Of Acorus a fragrant essence breathes—
> Nature's own incense shed to sanctify the wreaths.
>
> "O Rushes green!
> With blossoms wan or brown! And ye
> Sweet Flags, from whose scent-roots to me
>   Come thoughts of the Has Been,
> Ye are the fitting plants at eve to shed
> A vague mysterious perfume o'er the silent dead.
>
> "' Not so, not so!'
> A voice replies; ' for joy alone
> These Reeds and Rushes here are strewn!'
>   But I again cry, ' Lo!
> Joy's emblems here I fitly use, to prove
> That Life and Death alike spring from God's holy love.'"

My specimens were gathered near Warwick. Dr. Murray states that rice-paper is made from a Chinese species of Calamus. The Chinese bring it in their junks from Formosa. The pith is taken out and soaked in water, then sliced and pressed. Refuse pieces are used in making artificial flowers, and the sheets of six or ten inches in length are used for drawings of Chinese costumes, flowers, landscapes, &c.

The DUCKWEED order is the next in succession. They are all minute floating plants, covering the surface of ponds and stagnant ditches. They are often blamed for the unwholesome odour which arises from the ponds where they flourish; but this is exceedingly unjust, for in reality they purify the air to some extent, consuming a part of the noxious gas given forth by the stagnant water.

There are some ponds near Warwick where the Lesser and Gibbous Duckweeds (Lémna minor and L. gíbba), abound. The former has very small leaves; flat above and below, connected in floating masses; the latter has larger leaves, very thick, and egg-shaped underneath.

The Greater Duckweed (L. polyrhiza), has oval leaves, which are purple underneath. It is one of the scarcer species, but I had the good fortune to fish some out of the ditch that partly encircles Christ Church meadows, Oxford.

The other rare species is the Ivy-leaved Duckweed (L. trisúlca). Its leaves are lobed, and have stalks to them. We have specimens from near Bishopton in Durham.

The great use of Duckweeds, besides somewhat purifying the air, is to shelter insects, and thus provide both animal and vegetable food for the ducks, which gobble up the weed so greedily. The roots of these tiny plants consist in simple threads hanging from their leaves. The flowers are rarely found; they proceed from the side of the leaf, and consist in two stamens and one stigma. It is related of a celebrated botanist that his chief desire was to see the Duckweed in flower before his death.

The Pondweed order is composed chiefly of the extensive family which gives the name to the order. These plants have cellular stems, and either transparent or leathery leaves.

I think I have already mentioned an excursion which we all took to Hawksborough, with the intention of fossilising in the coal mines of that district. On arriving there our courage failed us, the ladies feared both the mines and the miners, and we were utterly at fault. There were some disgusting ponds, some heaps of coal rubbish, and a canal; these were all the resources that the place afforded. A party of us began to hunt about for plants, and having drawn the Spiked Milfoil from a pond, and being nearly smothered by the odour we evoked, we determined to try the canal. There was plenty of weed, and I hooked a quantity out.

My first piece of spoil was the Perfoliate Pondweed (Potamogéton perfoliátum). The leaves are not pierced by their stalk, as they should be to make the plant fully deserve its name, but they clasp the stem with their heart-shaped bases. They are brownish, and quite transparent, and when dry they resemble gold-beater's skin. The spike dips itself under water, and the few green flowers have four petals, four stamens, and four stigmas. All the Pondweed family are alike in the number of the parts of their flowers.

The Shining Pondweed (P. lúcens), also graced the canal; it has large lance-shaped leaves, also of a brownish hue, and very transparent, and its spike is large and rises above the water.

The common Chara (Chára vulgáris), I got likewise out of that productive canal.

A little later in July Edward was prowling about certain fishponds near Hawkhurst; there was a tangled mass of weed just beyond the reach of the longest stick he could find. Luckily some men were working in an adjoining field, and he borrowed a long rake from them and hauled out a quantity of the weed.

The Broad-leaved Pondweed (P. nátans), was there; its upper

leaves, leathery and oblong, floated on the surface, enabling the plant to rear its flowering-stalk erect. The leaves under water were narrow and transparent; the whole plant, leaves, stem and flowers were of a reddish-brown.

Entangled with it was the Grassy Pondweed (P. gramineum), with narrow green leaves, and flower-stalks from the forks; it lives and blooms under water.

In the pond at Hill Deverill, where the Reed-mace grows so regally, a pale green Pondweed, with ovate, sessile, crowded leaves, flourishes also. The spike is short, and pale green like the leaves; it bears about four flowers. It is called the "Close-leaved Pondweed" (P. dénsum).

I found a member of this family in a chalky pond on the Wiltshire Downs, and with much ado I got it. The plant was the Curled Pondweed (P. críspus). The leaves are lance-shaped, waved, and serrated; they are of an olive green, as is also the spike of flowers. The whole plant was under water.

The Smallest Pondweed (P. pusillum), is a most inoffensive little plant, with grass-shaped leaves and tiny spikes proceeding from the axils of the leaves. Fanny found it in the Loe Pool.

The Fennel-leaved species flourishes in the Clevedon ditches, its name expresses well its distinctive character (P. pectinátum).

Here is the Reddish Pondweed (P. ruféscens, *Plate XVII.*, *fig.* 7), I found it in a peat bog on Grinton Moor, in Swaledale. The whole plant is deeply tinged with red. The upper leaves are much broader than those under water.

There is a Various-leaved Pondweed, with stalked floating leaves on the upper, and sessile ones on the lower part of the stem; and there are a long-stalked species, and a Lanceolate Pondweed, both of which grow in pools among the highland mountains, but we have no specimens of them.

The Tassel-grass (Rúppia marítima), is in our collection. It is a graceful plant, bearing several grass-shaped leaves on its stem. The flowers are in a cluster and almost sessile, but their little footstalks lengthen fast, and ere the seed is ripe the cluster has become tassel-shaped. The main stalk also lengthens greatly, and coils itself round and round until it elevates the fructification out of the water. When the seeds are ripe the stem draws itself down and hides the whole cluster in the mud. The Tassel-grass grows in the salt-water ditches about Clevedon, and in the Loe Pool.

I have never found the Horned Pondweed (Zannichéllia palústris). It has a thread-like stem and grassy leaves, small axillary flowers; the male consisting of one stamen with no calyx or corolla, and the female of a calyx of one sepal, and four styles. The whole plant lives under water.

The Grasswrack (Zostéra marína), is common in the sea and salt ditches. Its tiny flowers seem to emerge from a sheath in the long grassy leaves, and consist in one style and two stigmas.

The extensive alliance of GLUMACEOUS plants is divided into two orders—1st, the Sedges; 2nd, the Grasses.

The Sedges differ from the Grasses in their stems, leaves, and flowers. The stem of Grasses is generally hollow and round; while that of Sedges is solid and angular. The overlapping leaves of Grass open on one side; but as the stem emerges from the leaf of Sedges, the base of the leaf forms a perfect cylinder. Grasses have chaffy calyces and corolla, but in the Sedge the male flower has only a bract, and the female a few bristles to serve the purpose of calyx and corolla. We cannot boast of the uses of this Sedge tribe.

The Bullrush and others of the tribe are used for making mats and the bottoms of chairs, and they are greatly esteemed for thatching.

The Pitsi, or Water-chestnut, which the Chinese grow in tanks and eat for dessert, is a member of this tribe.

The Cypérus of the Nile supplied the paper of ancient times. Pliny tells us that the plant was separated into thin slices by a fine point; these were brought into contact by their edges, transverse slips were then laid upon them, and then the whole was sprinkled with Nile water and submitted to pressure.

Dr. Murray says that if a portion of this Egyptian paper is held up to the light the ribs may be seen parallel, and crossing each other at right angles. The more modern reed-paper of Syracuse is made of the pith, not of the rind of the Reed.

The Bullrush of the Nile (Cypérus nilóticus), was the plant of which the ark was made in which the infant Moses was exposed. Boats were constructed of the stems of this Bullrush placed parallel and then transverse, and sails and cordage were made of the rind. Pliny says that the only boat they have in Abyssinia is formed of Reeds sewed and gathered, stem and stern, with a piece of Acacia as keel.

The long drooping stems of the Sedge family, with their waving leaves and bowed spikes tinged with purple, procured for them in old times the name of "Long Purples." Thus England's great poet speaks of them as forming a part of the last wreath of the unfortunate Ophelia.

> "There is a Willow grows aslant a brook,
> That shows his hoar leaves in the glassie stream;
> There with fantastic garlands did she come,
> Of Crow flowers, Nettles, Daisies, and Long Purples,
> And our cold maids do Dead Men's Fingers call them."

And Tennyson adopts the same name—

> "Round them blow, self-pleached and deep,
> Bramble Roses faint and pale,
> And Long Purples of the dale."

In the touching little poem of "The May Queen" the Bullrush is enumerated among the delights which will be past for her—

> "When the flowers come again, Mother, beneath the waning light,
> You'll never see me more in the long grey fields at night;
> When from the dry dark wood the summer airs blow cool
> On the Oat-grass, and the Sword-grass, and the Bullrush of the pool."

The Sedge Warbler and the Water-hen would be highly indignant if they heard the utility of the Sedges questioned. Among them they find a home and food, and there they bring up their little ones.

The first family in the Sedge tribe, that of the Bog Rush, has only one member.

The Black Bog Rush (Schœnus nígricans), has a round naked stalk, and a roundish head overtopped by two bracts. It grows in turfy bogs, and flowers in June.

The Beak Rushes occupy similar situations; there are two species, the White and the Brown; they have spikes containing a few flowers. The White Beak Rush has two stamens in each flower, and the Brown Beak Rush (Rhyncóspora fúsca), has three. These plants grow on Rudd Heath. I have gathered them there at the end of July.

The Cyperus family is the third in the Sedge tribe.

The Sweet Cyperus or Galingale (Cypérus lóngus), has many spikes united in a cluster, accompanied by two or three bracts. It grows two feet high, and is a rare ornament of marshes. The root is very aromatic, and Gerarde says, "It hath a most sweet and pleasant smell when it is broken."

The Brown Cyperus (C. fúscus), is a minute plant about four inches high. Its clusters of small spikes are brown, and accompanied by two bracts. It is a very rare plant; but a specimen gathered in Chelsea meadows has been given to me.

There is a Spanish Cypérus, with tuberous roots, which is eaten as salad. Quaint old Gerarde tells us that the people of Verona used to call it "Traci dulci," and that the vendors were wont to cry it in the streets. Dioscorides and Galen recommend it as medicine.

The Bullrush and Club Rush family is the fourth in the tribe; they form a numerous family, which need to be divided into groups.

First come those with solitary spikes.

The Scaly-stalked Club Rush (Scírpus cæspitósus), has scales at the base of the stems. Fanny found it near Clevedon.

The Floating Club Rush (S. flúitans), is only a few inches high, and floats on still water. This she gathered in Marazion Marsh; and the Chocolate-headed Club Rush (S. pauciflórus), has single spikes of brown flowers, and no scales at the base of its stems.

The second division have round stems and several spikes.

The tall Bullrush (S. lacústris), belongs to this division, and its compound panicle of spikes rises upon stems eight feet high. Edward has splendid specimens from his beloved Avon.

The Bristle-stalked Club Rush (S. setáceus), is a very slender plant; its stems and leaves are like threads. The spikes are few and small, and the whole plant does not attain a greater height than three or four inches. This specimen grew on a dry mud-bank at the mouth of the Loe Pool.

A specimen of the equally slender Savi's Club Rush (S. Sávii), was sent to me the other day from Sandown, in the Isle of Wight. Its mark of distinction is the roundness and roughness of the fruit.

There are a glaucous Club Rush (S. gláucus), inhabiting salt marshes; and a Round-clustered Club Rush (S. holoschœnus), a tenant of the seashore; and a Brown Club Rush (S. rúfus), with smooth fruit, which grows in marshes in the West. All these belong to one division.

The third division includes all that have triangular stems.

The Triangular Club Rush and the Blunt-edged Club Rush (S. triquétrus and S. carinátus), have no bracts with their cluster; the former lives on banks of rivers near the sea, and the latter on those of inland rivers.

The Sea Club Rush (S. marítimus), grows in salt marshes. Fanny found it in Lostwithiel, in Cornwall; and the Wood Club Rush (S. sylváticus), favours thick woods and swamps. I have gathered it in ditches near Clevedon. These two last have bracts in their panicles; the former is slender and glaucous, the latter strong in its habit, and with beautiful fresh green sword-shaped leaves.

The Spike Rushes are very simple plants, with leafless stems bearing a head of three or four florets at the summit.

The Many-stalked Spike Rush (Eleócharis multicáulis), grows

on the banks of the Clevedon ditches; its stems spread, and the plant grows in a matted cluster.

The Least Spike Rush (E. aciculáris), is a frail little plant, resembling the Many-stalked species, except that its stem is square; while that of both the other Spike Rushes is round, and it is very slender and minute. My specimen was sent me by a Scotch friend.

Edward has got the Creeping Spike Rush (E. palústris); it grows about Hawkhurst. The plant has no leaves, and it sticks up its rushy stems, with their insignificant spikes, on moist ground. The root is creeping.

I must take you to our Yorkshire or Scotch moors to show you the Cotton-grass in its beauty. The flowers are very unassuming, being arranged in spikes of a greyish hue, the stamens and anthers forming its only ornament; but in seed it is very beautiful, the tufts of snow-white cotton sprinkling the dark moor, and nodding in the wind.

The Hare's-tail Cotton-grass (Erióphorum vaginátum), is the most common species, growing in profusion amongst the Heath. The stem is triangular above, and round below, and bears one spike on its summit.

The Round-headed Cotton-grass (E. capitátum), has also a solitary spike; but its only habitat, according to Sir J. E. Smith, is on Ben Lawers.

The Broad-leaved Cotton-grass (E. polystáchion), has flat lance-shaped leaves, and a great number of small drooping spikes. I gathered it in fruit a year ago last May, in a shaking bog near Tiverton.

The common Cotton-grass (E. angustifólium, *Plate XVII.*, *fig.* 10), is the handsomest member of the family. The spikes droop as in the last species, but are much larger, and the leaves are very narrow. It abounds in marshy ground among the Yorkshire moors.

The Slender Cotton-grass (E. grácile), I found on the banks of the Swale above Melbecks, very early in the spring. The seed was evidently remaining from the last year, as its time of flowering is August. This is an alpine species, and could only have grown there by means of seeds washed down from the hills. A great quantity of the plant was there at the time I speak of, but a heavy flood occurred soon after, and I have since sought it in vain. It is a slender Grass with a round stem, and a somewhat drooping cluster of small spikes. Ossian speaks of these Cotton-grasses as "Cana-grass," and the Scotch adopt his term, calling it "Canna." He compares the complexion of his heroines to the "snow-white Cana." The down of the seed is sometimes collected for stuffing pillows.

The family of the true Sedges comes next. The stamens and stigmas are in different flowers, often in different spikes, and, in one species, on separate plants. The female flower consists of a capsule and a scale, and the male of a bract and three stamens; they are arranged in spikes or catkins.

The species with the male and female flowers on separate plants is called Creeping Sedge (Cárex dioíca); it is less than a span high, and very slender and smooth. The catkin is solitary in this species, as also in the Prickly Flea and Few-flowered Sedges, all of which frequent bogs in hilly country.

The Prickly Sedge (C. Davalliána), has round catkins, with the male and female flowers mingled; it grows rather above a span high, and is pretty common. I have found it about Richmond, and Malcolm has it from Hawkhurst.

The Oval-spiked Sedge (C. ovális), is of the same habit, but larger. I found it also near Richmond.

The Remote Sedge (C. remóta), is a very grassy plant, extremely slender, and with small catkins in the distant axils of the leaves. It was one of my prizes during an excursion to Brignall Bank and Rokeby.

The Sea Sedge (C. arenária), has half recumbent stems and oval catkins; the Soft Brown Sedge has dark-coloured catkins, and is tall and graceful; the Compound Prickly Sedge (C. vulpína), is a large plant with bright green leaves, and spreading erect catkins; and the Great Panicled Sedge (C. paniculáta), is a splendid ornament of ponds and ditches, growing three feet high, with broad handsome leaves and very large catkins. All these species we have from Yorkshire; they agree in the character of the mingled catkins. There are besides several similar species too numerous to describe singly. Of those with the male and female flowers on different catkins we have a great variety.

Fanny has the Dwarf Silvery Sedge (C. clandestína), measuring only four inches in height, from St. Vincent's Rocks, Bristol; and Edward has the Great Pendulous Sedge (C. pendula), from Hawkhurst. This is a noble plant, sometimes attaining the height of six feet, and with numerous stalked drooping catkins, each two or three inches long.

The Loose Pendulous Sedge (C. strigósa), is also his trophy from the same neighbourhood; it is shorter and more slender than the last-named, but resembles it in habit.

The Pendulous Wood Sedge (C. sylvática, *fig.* 8), I have from Easby Woods; its catkins are slender, loose, and drooping, and its leaves a pale yellow-green. The plant is seldom above a foot high.

The Bastard Cyperus Sedge (C. pseudo-cypérus, *Plate XVII.*, *fig.* 9), is also a Kentish plant; it is large and handsome, with drooping catkins. In size and habit it resembles the Great Pendulous Sedge, but the catkins are much thicker and shorter.

The Pale Sedge (C. palléscens), is very small and slender, but upright. Its oval catkins are few, and the lower ones pendulous. My specimen is from Wicliffe, near Ripon.

The Tawny Sedge (C. fúlva), has rust-coloured catkins, and a rough stem. The Green-ribbed Sedge resembles it, and has a very dark catkin, drooping and elegant; and the Vernal Sedge (C. præcox), is a bright little plant, a few inches high, with upright spikes. These three flourish in the vicinity of Richmond.

The Glaucous Sedge (C. recúrva), is so named from the

colour of its leaves. It is a span high, and has pendulous purple catkins.

The Great Common Sedge (C. ripária), has its catkins upright when in flower; they are tinged with purple, the leaves are glaucous, and the plant two feet high, bowed and graceful.

The Flagon Sedge (C. ampullácea), has its spikes arranged with great precision like the ears of Indian Corn; it is about one foot high. The Hairy Sedge (C. hírta), resembles it, except that its herbage is exceedingly hairy. These I found near Bishopton, in Durham.

There are many other species of Sedge, but it is impossible to give a description of each without becoming tedious.

CHAPTER XVIII.

GRAMINÁCEÆ.

> "'Let the earth
> Put forth the verdant Grass, herb yielding seed,
> And fruit tree yielding fruit after his kind,
> Whose seed is in itself upon the earth.'
> He scarce had said, when the bare earth, till then
> Desert and void, unsightly, unadorned,
> Brought forth the tender Grass, whose verdure clad
> Her universal face with pleasant green."
>
> MILTON.

A GRASS is the simplest form of a perfect plant. A thin stem, clothed with alternate leaves, constituting sheaths to guard the fast-growing buds, a few flowers collected at the end of the branches, a very few stamens, and a single seed, constitute the whole plant. Yet, simple as these parts are, the various genera are very clearly distinguished, so that this family is one of the easiest to understand. That part of the flower which is generally called calyx is termed the *glume* in Grasses, and paleæ stand for petals. The bristle often accompanying the flower of Grass is called an *awn;* the stamens are generally three in number, and the styles two, which are often beautifully feathered. The roots are in all cases fibrous. The stem is generally cylindrical, though sometimes compressed, varying in length from two inches, as in the Sand Cat's-tail, to eighty feet, as in the Bamboo. The leaves are one to each knot in the stem, sheathing at the base, and slit on one side. The male and female flowers are generally on one spike, nearly always on one plant. The glumes are generally two to each spikelet, the outer being the largest; but sometimes there is only one, and there are cases where both are absent. The embryo lies at the side of the lower part of the seed.

The Sweet Vernal-grass (Anthoxánthum odorátum), is the only British representative of the first family in this extensive tribe. It is distinguished from all other families by having two stamens only, and this throws it into a different class from the other Grasses in the Linnæan system. Our Sweet Vernal-grass is a small annual plant, bearing its flowers in short heads, not very compact, and broader at the bottom than at the top; it has two sharp-pointed unequal glumes, and two dark brown hairy paleæ, with awns on their back. This Grass is of little importance to the farmer, its nutritive qualities being very small; but it is esteemed for its aroma, which imparts the well-known fragrance to new-mown hay, and it is interesting as being one of the earliest-flowering Grasses.

Our one British Mat-grass is very common on the Yorkshire moors, and I see that Fanny has specimens from near the Cheese Wring, in Cornwall. The leaves are narrow and plentiful, the spike about a span high, with one row of florets; their anthers generally hanging to one side. The glumes and paleæ are violet colour, and the anthers sulphur; the foliage is dark green. This Grass forms thick tufts upon moist waste land; it is most fitly named Mat-grass (Nárdus stricta, *fig*. 14). There is only one stigma to each flower.

The Canary-grass is familiar on account of its furnishing bird seed. The species thus useful flowers in a round head.

In both the common Canary-grass and in the Reed Canary-grass there are two glumes and three or four paleæ.

The so-called common Canary-grass (Pháláris canariénsis, *fig*. 1), is a rare plant, but Fanny has met with specimens from Potato fields near East Looe, Cornwall. It is a native of the Canary Islands.

The Reed Canary-grass (P. arundinácea), I have found frequently; it grows by the side of the Ouse, near Little Ouseburn, in the York district. Unlike its brother, its flowers are arranged in a handsome panicle.

A variety of this plant, with yellow-striped leaves, is the

Ribbon-grass of our gardens. It is a useful Grass in its natural habitat, for it binds together the loose sand and earth by the river's side: thus forming a natural barrier against inundations. Both these Grasses are tall; the one I have just described growing four feet high, and the Canaryseed attaining the height of three feet.

The Cat's-tail-grasses have two pointed glumes concealing the blunt insignificant paleæ.

The common Cat's-tail (Phléum praténse), is frequent in pastures and meadows. Its flowers are packed into a close cylindrical spike, the stem rises to the height of a foot or a foot and a half. It is not a very profitable meadow Grass, being too thin and wiry, but it is excellent for permanent pasture.

There are an Alpine Cat's-tail and a Rough Cat's-tail without awns, and a Purple-stalked Cat's-tail; but all these are rare Grasses, and of no account to the farmer.

The Sea Cat's-tail (P. arenárium), is a minute Grass with a tapering spike, hairy foliage, and glaucous tint. Fanny's specimen is from the shore at Weston-super-Mare.

The Foxtail-grasses are characterised by two closely-drawn glumes, one palea, with an awn rising from its base, and the two styles combined at the bottom. The florets are arranged in cylindrical spikes, in general appearance resembling the Cat's-tail.

The Meadow Foxtail (Alopecúrus praténsis), is very valuable in meadows, being exceedingly early, and very abundant in its growth.

The Slender Foxtail (A. agréstis), flowers later in the season, and frequents corn fields and roadsides; the stem is roughish, and the spike very tapering. Both these species grow three feet high.

The Floating Foxtail (A. geniculátus), has its stem bent at the joints; it is a rough Grass, and of much lower growth than the two already mentioned, though the length of its stem depends a good deal on the depth of the water on which it floats. It is a common Grass on the margin of pools.

There are an Alpine Foxtail (A. alpínus), with very small spikes, and a Bulbous Foxtail (A. bulbósus), inhabiting salt marshes; but we have no specimens of them.

There is a little Grass called the Early Knappia, which frequents seashores; it has two glumes and two unequal hairy paleæ. It flowers in March and April, and only attains the height of three inches.

The Beard-grasses also are maritime plants. The annual species has awns thrice the length of the glumes; it grows among sand. The perennial species has shorter awns, scarcely rising higher than the glumes; it favours salt marshes. Both have their florets arranged in close panicles, and grow about a foot high. They bloom in August. The annual Beard-grass was sent to me from the coast of Norfolk; the perennial Fanny has from Clevedon (Polypógon littorális).

The British Millet-grasses are unimportant plants, though the extreme elegance of the Spreading Millet gives it an indisputable claim to beauty (Mílium effúsum, *Plate XVIII.*). In this family the florets have two close glumes enclosing the two paleæ, which harden and form a coat for the seed.

The Spreading Millet grows four or five feet high; it has broad pale green leaves, and a large scattered panicle of tiny green flowers. It grows abundantly in woods and shady places, several stems rising from one root. The woods in Swaledale are full of it.

The Panick Millet (M. lendígerum), has its flowers in a dense spiked panicle, and the paleæ are awned. It grows in inundated fields, and is rare.

The Bent-grass family is characterised by two unequal glumes enclosing a single floret; the paleæ are transparent, the larger having sometimes an awn at its base. These are elegant Grasses with fine foliage and light panicles of flowers.

The Silky Bent-grass (Agróstis spica-vénti), has its panicle waving to one side; it grows in sandy fields.

The Brown Bent-grass (A. canina), is common on roadsides,

by field paths, and in moist places; the brown hue of its florets distinguishes it. These, and the Bristle-leaved Bent-grass, have awns.

The Fine Bent-grass (A. vulgáris), is a familiar plant in meadows and pastures, its very slender shining purple panicle crowning a stem less than a foot high.

The Marsh Bent-grass (A. álba), grows abundantly in swampy ground; it is the Fiorin-grass about which a great fuss has been made, it being argued that it would make super-excellent marsh herbage. It grows to a great size and richness in Ireland; but it does not attain any excellence in our English marshes. It is called by country people "Quicks,"* and is a troublesome weed in corn land. These two Bent-grasses have no awns.

The Dog's-tooth-grass (Cýnodon dáctylon), has two lance-shaped, acute, spreading glumes, and two very unequal compressed paleæ. The florets are arranged in loose spikes, five spikelets in a cluster; the leaves are sharp-pointed and hairy. It frequents our seashores, flowering in July and August. Fanny has it from Cornwall. This Grass is called "Doob" in India, and held sacred by the Brahmins. The European settlers make their lawns of it, sending the natives to collect the plants from the plains.

The Finger-grasses resemble the Dog's-tooth-grass in the arrangement of their flowers, but they have sometimes three glumes.

The Hairy Finger-grass (Digitária sanguinális), grows in fields, but rarely; our fortunate friend has met with it also in Cornwall.

The Smooth Finger-grass (D. humifúsa), is a smaller plant, and is peculiar to the eastern coast.

The Panick-grasses form an important family. To it belong the true Millets of India, Arabia, and the south of Europe— Grasses which, for their utility for human food, deserve to be ranked among the corn group.

There are gigantic Panick-grasses in Brazil, with delicate and tender foliage, most valuable for cattle; and the Guinea-grass, or Great Panick, forms the most profitable pasturage in Jamaica.

Our British Panick-grasses occur rarely, and in small quantities; they have two or three florets enclosed between each pair of glumes, one of which is neuter. The glumes are ribbed.

The Rough Panick (Pánicum verticillátum), has smooth-jointed stems, and the glumes keeled.

The Green Panick (P. víride, *fig.* 4), of which I have received a specimen from Norfolk, has erect stems, and a crowded spiked panicle; and the Loose Panick-grass (P. crus-gálli), has its flowers to one side, the panicle branched, and the leaves lance-shaped and harsh.

The Hair-grass family have also two or three florets enclosed between each pair of glumes; the paleæ continue unchanged until the seed is ripe.

The Crested Hair-grass (Áira cristáta), grows only eight inches high; its panicle is lanceolate and downy. The Water Hair-grass (A. aquatica), has a spreading panicle, and its glumes are abrupt and notched. These two have no awns.

The Tufted Hair-grass (A. cœspitósa), is an elegant plant, raising its large light panicles of glossy florets three feet high. It grows in mat-like masses in woods and hedges; the florets have one awn from the bottom of the outer glume.

The Wavy Hair-grass (A. flexuósa), is smaller; its panicle contains fewer florets, and the awn rises from the middle of the outer glume. We have these Grasses in great abundance about our woods and moors in Swaledale; and Fanny has the Waved species from the Cheese Wring, Cornwall.

The Silver Hair-grass (A. caryophýllea), I gathered at Plumpton Rocks, near Harrogate; the cluster is less dense even than that of the last-named species, and the flowers have a silvery hue; the awn is twice as long as the glumes.

The Grey Hair-grass (A. canéscens), was sent to me from Ventnor, where it grows on the seashore; its panicle is much more dense than those of its brethren, and it is less attractive in consequence.

Edward has the Early Hair-grass (A. præcox), from Hawkhurst. It is a very neat little plant, growing one or two inches high, and flowering in May; its panicle has very few florets.

The Soft-grasses do not form first-rate pasturage, as all their foliage is more or less hairy. Like several of the families recently mentioned, that of the Soft-grass has two or three florets enclosed between each pair of glumes. The paleæ form the coat of the seed, and are awned. The glumes are keeled. Species of Soft-grass inhabiting India, Arabia, and the Cape, are used as Millet, and called by the same name, as well as the Panicks.

The Meadow Soft-grass (Hólcus lanátus), is very woolly. The lower floret is perfect, and without an awn; the upper is awned. It grows freely in meadows and pastures, and its dense panicle of downy florets, slightly tinged with pink, are very familiar.

The Creeping Soft-grass (H. móllis), flowers a fortnight earlier than the Meadow species, coming into bloom in the middle of June. It has fewer flowers in its panicle, and the upper floret has a very prominent awn. These two have a great resemblance to one another. I have gathered both frequently in Swaledale, and Edward has them from Hawkhurst.

The Oat-like Soft-grass (H. avenáceus), is quite as common as the others, and no member of the tribe can exceed it in beauty. It bears no outward resemblance to the other two Soft-grasses, which do not much exceed a foot in height. The Oat-like species grows in hedges and waste places, and sometimes among corn. I have often seen it five or six feet in height. Its leaves are broad, deep green, and rough. Its panicle is spread and tapering; the florets large, with unequal glumes, and a prominent awn to the lowest floret. The stamens are long, and the anthers deep purple. It is a very useless Grass, though so handsome.

The Holy-grass (Hieróchloe boreális), I have never seen, except in the Botanical Gardens at Edinburgh. It is an elegant grass, with pale green foliage, and scattered panicles of florets, three to each spikelet. The centre flower is perfect, and has two stamens; those at either side are barren, and have three stamens. It inhabits lofty glens in Scotland.

The Melic-grasses are among the most elegant of the tribe, but, as is often the case with the beautiful Grasses, they are of no great service. They have one or two florets between the glumes.

The Wood Melic-grass (Mélica uniflóra), has the panicle slightly branched; it is turned to one side, and somewhat drooping. The barren florets are stalked, and the fertile are seated; they are of a beautiful violet hue. The foliage is broad, and of a delicate pale green. This Grass abounds in woods everywhere.

The Mountain Melic (M. nútans), is more scarce. I have only found it in Mackershaw Woods, near Ripon. The flowers are arranged in a drooping spike; the paleæ have no awns.

The Purple Melic (M. cœrúlea), has narrower foliage than the others, and it has a slightly glaucous hue. The florets are violet-coloured, much smaller than those of its brethren, and more numerous. It grows abundantly on damp ground on the Yorkshire Moors.

The Blue Moor-grass (Sesléria cœrúlea), inhabits similar places where the soil is chalky. It has two or three perfect florets on each spikelet, and the whole are arranged in an oval cluster or head. The glumes have a blue tinge, and the foliage is narrow.

We now come to the great Meadow-grass family, which is characterised by having several flowers in each spikelet, no awns, and the seed loose. This family is now divided into two—the Sweet-grasses and the Meadow-grasses, the former having the paleæ simple, the latter having them keeled.

We have the Reedy Sweet-grass (Glycéria aquática), which Edward found on the margin of the Avon, near Leamington, and also in a pond opposite the Hall House, Hawkhurst. It is a tall Grass, with stiff sword-shaped leaves, and an erect panicle of small spikelets. The florets are all blunt.

The Floating Sweet-grass (G. flúitans), is to be found everywhere on the margin of ponds. Its panicle is oblong and branched, and its lance-shaped spikelets are close-pressed, and contain numerous florets, the paleæ of which are seven-ribbed.

Fanny has the Reflexed Sweet-grass (G. dístans), and also the Hard Sweet-grass. The former grow on salt-marshy ground beside the Looe River. It is partly recumbent, and its panicles are stiff. It grows about twelve inches in length, and thus varies exceedingly from the above-mentioned members of its family, the Reedy Sweet-grass attaining the height of four feet, and the Floating of three feet.

The Hard Sweet-grass (G. rígida), is a very small plant, inhabiting the tops of walls and sandy places. Its stems are wiry, and its panicle is formed of close-pressed spikelets, all of which lean to one side. This specimen is from the Looe Cliffs.

There are a Procumbent Sweet-grass (G. procúmbens), with a stiff panicle, which inhabits seashores; and a Sea Sweet-grass (G. marítima), the panicle of which is erect and branched, and very close; but we have not got specimens of them.

The true Meadow-grasses come next. Their botanical name, Poa, is the Greek word for Grass of any kind. The plants of this family have a pair of glumes to each spikelet of many florets, and the paleæ are membranous at the point.

The Rough Meadow-grass and the Smooth-stalked Meadow-grass (Poa triviális, and P. praténsis), are exceedingly useful for meadow crops. They grow eighteen inches high, with full-branched panicles of small spikelets, very often tinged with purple. The roughness, or otherwise, of the stem and leaves is a sufficient mark of distinction between them.

The Annual Meadow-grass (P. ánnua), is the common weed of our gardens, infesting the untrodden pavement as well as the green lane, and flowering from March to November.

The Alpine and Glaucous Meadow-grasses were sent to me from Scotland. The former has a denser panicle than most of its family, and the glaucous, broadish foliage of the latter distinguishes it.

The Wood Meadow-grass (P. nemorális), has a graceful slender panicle, and only about three flowers in each spikelet. Both spikelets and leaves are of a pale green. It grows in the Easby woods.

There are a Flat-stalked Meadow-grass, the distinctive mark of which is expressed in its name; and a Bulbous Meadow-grass, with a zigzag panicle and four-flowered spikelets. This grows on the seashore.

The Heath-grass (Triodia decúmbens), used to be counted among the Poas, but its round paleæ and concave glumes are marked enough to allow it a genus to itself. It is a rigid plant growing in tufts, the stems often leaning towards the ground at an acute angle. The panicle is very little branched; there are four florets in each spike. It grows in swampy places on our moors in Yorkshire, and flowers in July.

The Quaking-grass is a familiar object to people of all ages. About Nantwich, in Cheshire, it used to be called "Quakers and Shakers," as old Gerarde tells us. With us it is often called "Trembling-grass."

The common Quaking-grass (Bríza média), is known to us all. It has oval spikelets of seven flowers, and the glumes are shorter than the florets.

The Small Quaking-grass (B. mínor), is a rare plant. My specimen came from Devonshire. It has triangular spikelets, with seven flowers in each; the glumes are longer than the florets. The panicle is dark green.

Y

Of the Cock's-foot-grass family (Dáctylis glomeráta, *Plate XVIII., fig.* 7), we have only one British species. It is characterised by two sharp-pointed glumes, which are keeled and enclose from three to six florets. The panicle is distantly branched, the lowest branch being long, and standing at right angles with the stem, like the claw on a cock's foot. This is a valuable Grass, because it endures drought so well, and on this account it is such a favourite in Norfolk. Although a coarse Grass, it is much liked by cattle, and it becomes finer by cultivation. It grows from two to three feet high.

The Cord-grass (Spartina strícta, *fig.* 8), is the only member of its family. It has two lance-shaped glumes, and two unequal paleæ; there is only one floret between each pair of glumes. The flowers are arranged in spikes, two or three spikes together. The leaves are rolled-in, ribbed, and pointed. It is an inhabitant of muddy salt marshes.

The Dog's-tail-grasses are two in number. They have both glumes and paleæ awned, and the florets are put on to the spike in pairs.

The Crested Dog's-tail (Cynosúrus cristátus, *fig.* 12), has the spike straight, and all the florets turned to the front, so that the back of the stem is left naked. It is a valuable pasture Grass.

The Rough Dog's-tail (C. echinátus), has a compound spike, and the awns on the paleæ are as long as the glumes. It grows the most freely in Jersey.

The family of the Fescue-grass numbers several members.

The Sheep's Fescue (Festúca ovína), is a slender Grass, with a quantity of fine narrow foliage. Sheep are very fond of it, and it is said that they never enjoy hills where it is not found. Its stem is square, and the panicle is rather close, and leans to one side.

The Hard Fescue (F. duriúscula), has also a one-sided panicle, but it is larger and more spread. The florets are longer than the awns, and the stem is flat. This Grass stands drought the best of any.

The Creeping Fescue (F. rúbra), resembles the Hard Fescue, and, like it, is a good pasture Grass. It is distinguished by its leaves being glaucous, and hairy on the upper surface.

The Sheep's Fescue and the Hard Fescue abound in hilly pastures in Swaledale. The Creeping Fescue Fanny found about the Looe River and harbour.

The Barren Fescue (F. bromoídes), is abundant on walls and sandy places. Edward has it from Hawkhurst. The panicle is narrow and erect, and the florets tapering, with long awns. Its stem only reaches a few inches in height.

The Wall Fescue (F. myúrus), resembles this, but has a more leafy stem.

The Tall Fescue has a compound drooping panicle, and from three to six flowers in each spikelet. It grows four or five feet high.

The Meadow Fescue (F. praténsis, *fig.* 9), is a valuable Grass for moist land. Its panicle is nearly upright, branched and spreading, and it attains the height of two feet.

The Spiked Fescue (F. loliácea), has its flowers in alternate spikelets, seated on the stem. It resembles the Rye-grass, but is distinguished by having two glumes, the other being only possessed of one.

The Wood Fescue (F. sylvática), is an elegant Grass, flowering in a drooping spike; the spikelets, which are placed alternately on the stem, being long and narrow, and adorned with long awns. The leaves are pale and hairy. It grows about two feet high, and adorns woods and hedges.

The Spiked Heath Fescue (F. pinnáta), resembles the Wood Fescue, but it has shorter awns. The foliage is smooth, and the height does not exceed eighteen inches.

All the Fescues have many-flowered spikelets. The lower paleæ are not awned as in the Brome-grasses, nor blunt as in the Poas, but they terminate in a hard sharp point. The glumes are acute and very unequal.

In the Brome-grasses (Brómus gigánteus, *fig.* 10), the paleæ are awned at the back, and the glumes are awnless. The spikelets contain several florets. They are of no use to the farmer.

The Smooth Rye Brome-grass (B. secalínus), has a spreading panicle, and ovate spikelets containing about ten flowers each. The leaves are slightly hairy, and the awns are shorter than the glumes.

The Downy Rye Brome-grass (B. velutínus), has a simple spreading panicle, and its spikelets contain from ten to fifteen flowers. The awns are the length of the glumes, and the whole plant is very downy.

The Smooth Brome-grass (B. racemósus), is also somewhat downy, but is taller and more slender than the two last species.

The Field Brome-grass (B. arvensis), has more slender spikelets, and its panicle rather droops. The leaves are hairy, but the stem is smooth.

The Upright Brome-grass (B. eréctus), is distinguished by its narrower leaves.

The Rough Brome-grass (B. ásper), has a much-branched, drooping panicle, and very large woolly leaves. It grows in moist woods, and attains a height of from four to six feet. Its spikelets are long and narrow, and the awns are shorter than the glumes.

The Barren Brome-grass (B. stérilis), inhabits sandy banks and waste places. Its panicle is but little branched, and its spikelets, containing eight flowers, are narrow lance-shaped. Eighteen inches is its usual height.

The Annual Brome-grass (B. diándrus), is a small plant, less than a foot high, with an upright simple panicle of conspicuous spikelets, from seven to eleven in number. The awns are very long. Fanny has a specimen of this Grass from Clifton.

The Barren, Rough, and Downy Rye Brome-grasses are in

my Yorkshire collection; and Edward has found the Field, Smooth, and Soft species about Hawkhurst.

The Oat-grasses are distinguished by their lax panicles, two loose membranous glumes, and small number of florets, each of which has one palea armed with a strong-twisted awn.

The flowers of the Wild Oat (Avéna fátua), so nearly resemble a fly that they are used instead of artificial flies for fishing. Its panicle is erect, and the spikelets pendulous.

The Narrow-leaved Oat (A. praténsis), is a rare inhabitant of chalky pastures; its branches are short and simple, and its florets about five in a spikelet.

The Alpine Oat (A. alpína), has an erect panicle also, but fewer flowers in the spikelets, and the stalk of each is bearded under the cluster.

Two of this group which used to be reckoned in this family are exalted by Persoon into a new genus, and called Trisetum, or "three-seeded." They have three flowers in each spikelet, the glumes have a membranous keel, and the lower palea has two bristles and a tender flexible awn above the middle of its back.

The Golden Oat (A. flavéscens), the first member of the new family, has a much-branched panicle with erect spikelets; it has a brilliant golden hue, and flourishes in rich pastures. It does not thrive well alone, but needs to be mingled with other Grasses. There is a bitter principle in it which adds to its value. Sheep are very fond of it.

The Downy Oat-grass (A. pubéscens), has also an erect spreading panicle, but its leaves are woolly; and it is accounted a second-rate meadow Grass, generally characterising poor land; but it greatly improves by cultivation.

The Darnel family have many-flowered spikelets, which are seated alternately on the stem, forming a narrow spike. There are one glume and two herbaceous paleæ. The two styles are very short, and the stigmas are feathery.

The Perennial Darnel, or Rye-grass (Lólium perénne), has the glumes shorter than the florets, the paleæ very slightly awned, and the foliage dark green and smooth. It is a valuable Grass both for meadows and pastures.

The Bearded Darnel (L. temuléntum, *fig.* 11), is distinguished by having long glumes, and the awns longer than the paleæ; it grows to the height of two feet, and its leaves are rough. The seeds of this Grass have a narcotic and deleterious quality, which produces intoxication in beasts and birds, and brings on convulsions. Christian says that when made into bread and eaten, the effect is giddiness, delirium, and paralysis. A farmer near Poictiers killed himself by insisting upon eating Darnel-bread. His wife and servant, who had eaten of the same at his command, escaped with only severe sickness. Old Gerarde says, "The new bread wherein Darnel is causeth drunkenness. Darnel hurteth the eies and maketh them dim, if it happen in corne either for bread or drink; which thing Ovid in his third booke hath mentioned, and hereupon it seemeth that the old proverbe came, that such as are dim-sighted should be said to eate of Darnel."

The Short-awned Darnel is a grass of little account, the paleæ little or at all awned, and the spikelets the length of the glumes; it is distinguished from the Bearded Darnel by its smaller size, and the smoothness of its stem.

The Hare's-tail-grass (Lagúrus ovátus), is a pretty curious plant; its head is composed of single florets contained within fringed glumes, each of which is surmounted by a feathery awn. Our specimen is the gift of Mr. Ward, of Richmond, who cultivates it in his scientific garden. It is a native of Guernsey.

The Reed family are characterised by having from three to five flowers in each spike, the two glumes being sharp-pointed and channelled, keeled, nearly equal, and as long as the florets. There are two membranous paleæ, the lowermost split at the end, and the split slightly haired. I suppose one of these

Reeds is the fabled plant into which Syrinx turned when she fled from Pan, and of which he made his instrument of music.

> "He filled his arms with Reeds new rising on the place;
> And while he sighs his ill success to find,
> The tender canes were shaken by the wind,
> And breathed a mournful air unheard before,
> That much surprising Pan, yet pleased him more.
> He formed the Reeds proportioned as they are,
> Unequal in their length, and waxed with care;
> They still retain the name of his ungrateful fair."

The Common Reed is as likely as any to be the transformed heroine of the story, as it inhabits the margins of rivers. It is a handsome plant with five florets to the spike, and a loose panicle. I have it from my favourite pond at Hill Deverill, where it grows above six feet in height. The tenant who farms the adjacent land mows these Reeds from a boat, and sells them for thatching purposes (Arundo phragmites).

The Wood Reed (A. epijégos), is a smaller plant, with an erect close panicle and long awns.

The Small Reed (A. calamagróstis), has slender grassy leaves, and a loose panicle. Both these species inhabit woods.

The Close Reed (A. strícta), is a highland species; it has only one flower in a spikelet.

The Cypress Reed (A. dónax), is a foreign member of this group; it was early used for walking-sticks, as Gerarde tells us, "This great sort of Reeds or Canes hath no particular description to answer your expectations; for that as yet there is not any man which hath written thereof, especially the manner of growing them, so that it shall suffice that ye do know that this great Cane is used, especially in Constantinople and thereabouts, of aged and wealthy citizens, and also noblemen and such great personages, to make their walking-staves of, carving them at the top with sundry scutcheons and pretty toies and imagerie for the beautifying of them."

The allied family, that of the Sea Reed (Ammóphila arundinácea), has one important British specimen; it has but one floret to each pair of glumes, and the flowers are arranged in a spike. It grows from two to three feet high; it is a rigid plant with bluish rolled-up leaves. It is a most useful Grass, piercing the sandbanks on the seashore with its roots, and converting them into natural barriers against the encroachments of the waters. On this subject Dr. Murray writes, "The very existence of a kingdom depends on Sea Reeds, Sedges, and kindred plants. These form the defence of the dykes of Holland, and prevent not only the invasion of the sea, but the advance of the drift sand on the fertile soil. When in a solitary walk by the seaside, we have heard in a still night the sentinel from the rampart repeat the watchword, 'All's well,' we have turned instinctively to the 'Sea Mat-weed' on the shore mantling the beech—the sentinel of Providence that forbids the approach of the waves, as if commissioned to say to them, 'Hitherto and no farther,' and 'Here shall thy proud waves be stayed,' and have then responded 'All's well.'"

The Sea Hard-grass (Rottbóllia incurváta), is a curious plant, with Rush-like branched stems, from the sides of which the little florets emerge. It only grows a few inches high, and inhabits salt marshes.

The Lyme-grass family have two glumes enclosing two or more florets. The flowers are arranged in spikes.

The Sea Lyme-grass (Élymus arenáreus), performs the same useful office as the Seá Reed, binding together the loose sand of the seashore. The large leaves are sharp-pointed.

The Pendulous Lyme-grass (E. geniculátus), is a very elegant plant; its spike is loose and bent downwards, and it is very tall and slender.

The Wood Lyme-grass (E. europǽus), has two rough florets between each pair of glumes, and its leaves are soft. Our specimens were sent from Lincolnshire and Norfolk.

We have now gone through the British members of the Grass tribe, with the exception of those that bear close affinity to the Cereals or corn Grasses; they are a humble race compared with the Pampas-grass of America with its towering stems, or to the Tussac-grass of the Falkland Islands. This Grass forms balls round its roots, five or six feet high, and from the top of these rise the stems and leaves; the latter drooping to the extent of seven feet. The Grass grows in tufts some distance apart from one another. There are no trees in these islands, nor any bushes larger than our Furze, so the Tussac is much valued. It fattens cattle very rapidly; the roots are sweet and good for food, resembling fine Cabbage in taste when boiled; and building cement is formed of the chopped Grass mingled with clay. This tribe of plants is interesting as the first green things which God created. "And God said, Let the earth bring forth Grass." "He causeth Grass to grow for the cattle." Grass is often used in Scripture as a type of the evanescence of all human things. "All flesh is Grass, and all the goodliness thereof as the flower of the field; the Grass withereth, the flower fadeth, but the word of our God shall stand for ever." And more especially of the short continuance of the wicked: "When the wicked spring as the Grass, and all the workers of wickedness do flourish, it is that they shall be destroyed." "The Lord shall smite Israel like a Reed shaken by the wind." But, although such solemn truths are associated with the Grass of the field, yet its presence is ever a proof of the kind providence of God, and as such must both assure and cheer us. An American poet has interpreted pleasantly, "The voice of the Grass."

> "Here I come creeping, creeping everywhere;
> By the dusty roadside,
> On the sunny hillside,
> Close by the noisy brook,
> In every shady nook,
> I come creeping, creeping everywhere.
>
> "Here I come creeping, creeping everywhere;
> All round the open door,
> Where sit the aged poor,
> Here, where the children play
> In the bright and merry May,
> I come creeping, creeping everywhere.'

> "Here I come creeping, creeping everywhere;
>   In the noisy city street
>   My pleasant face you'll meet,
>   Cheering the sick at heart,
>   Toiling his busy part,
> Silently creeping, creeping everywhere.
>
> "Here I come creeping, creeping everywhere,
>   More welcome than the flowers
>   In summer's pleasant hours;
>   The gentle cow is glad,
>   And the merry bird not sad
> To see me creeping, creeping everywhere.
>
> "Here I come creeping, creeping everywhere;
>   When you're numbered with the dead,
>   In your still and narrow bed,
>   In the happy spring I'll come,
>   And deck your silent home,
> Creeping silently, creeping everywhere.
>
> "Here I come creeping, creeping everywhere;
>   My humble song of praise
>   Most gratefully I raise
>   To Him, at whose command
>   I beautify the land,
> Creeping silently, creeping everywhere."

And James Ballantine draws a lesson of trust from the daily nourishment of the Grasses.

> "Confide ye aye in Providence, for Providence is kind,
> And bear ye a' life's changes wi' a calm and tranquil mind;
> Tho' pressed and hemm'd on every side hae faith and ye'll win thro',
> For ilka blade o' Grass keps its ain drap o' dew."

The various families of the corn Grasses may each be divided into two parts—1st, the eatable; 2nd, the field species.

The Wheat family needs to be twice divided, and the groups thus formed are called Wheats, Spelts, and Grasses. The botanical name of this family, Triticum, is derived from a word signifying *rubbed*, in allusion to the grinding of the grain.

Of true edible Wheats there are many varieties. The Hard Wheat has a compound spike, and grows very luxuriantly in Egypt; the Polish Wheat has long chaff and hard grains; the Red and White Wheat vary in different soils, the straw changing colour before the seed, so that we have Red Wheat with white straw, &c. All the true Wheats have the seed free from the chaff. Scientific men have named a district in Asiatic Russia as the native country of the Wheat, but the fact seems scarcely established; and the recent discovery of a similar plant by Stuart when crossing the interior of New South Wales raises a question whether its home be not there. It has been a familiar plant in cultivation in most countries and in remote periods, but nowhere has it become naturalised through cultivation.

Both the Egyptians and the Jews used Wheat. At the time of the Exodus "the Wheat and the Rye were not smitten, for they were not grown up." And David, lamenting the instability of his people, says, "He should have fed them also with the finest of the Wheat." Theophrastus and Pliny make frequent mention of Wheat, for it was used among the Greeks and Romans. It flourishes in the temperate regions, and is to them what Maize and Rice are to tropical countries. Wheat contains the largest portion of gluten of any of the cereals; its value as food is very great, forming the chief strength of the "staff of life." The straw serves for food for cattle, and is made into bonnets, hats, &c.

The Spelts are distinguished by the chaff adhering to the grain so strongly as only to be separated in the mill. Spelt has a parallel compressed spike, while Wheat has a four-cornered spike.

Couch-grass (Triticum répens, *fig.* 14), is the principal species of the third division of the Triticums. Though stigmatised as a corn-weed, and as such rooted-out and burned, this Grass has some valuable qualities; its creeping root is exceedingly nutritious when boiled, and in times of famine our forefathers were glad to be indebted to it for subsistence. The boiled roots of Couch-grass make excellent food for pigs. It is a tall Grass with five-flowered spikelets and flat leaves.

The Sea Wheat-grass (T. júnceum), works along with the

Lyme-grass and Sea Reed in constructing sea-barriers; its florets are awnless, and its leaves rolled-in, sharp-pointed, and glaucous.

The Fibrous-rooted Wheat-grass (T. canínum), has handsome awns, a straight slender green stalk, and roughish leaves; it is a nutritious Grass, and will grow freely on even poor land.

All the Wheat family have solitary spikelets, two glumes, enclosing many flowers, and two paleæ, the external one being pointed, the internal toothed. Our forefathers used to cut the Wheat half-way between the root and the ear that it might occupy less room in the barn, and the stubble was mown afterwards. We have too great a respect for the straw to follow this style of reaping. Wheat is subject to various diseases which discolour and spoil the grain: these are called Burnt-ear, Smut, and Ergot. It has also enemies from without: the ear-cocles are eel-like insects which eat into the ear, and the wheat-midge deposits its eggs in the germ and prevents the filling of the seed.

The Barley family (Hórdeum), is distinguished from the Wheat by its spikelets having only one perfect floret in each, and by the glumes being one-sided and bearded. The members of this family inhabit both the Old and the New World. America has eight different species.

The cereal species of Barley can bear great extremes of temperature. In Spain and Barberry two crops of Barley are grown in one year; and when "the Flax and the Barley were smitten, for the Barley was in the ear," it was the early crop, being the month of March. There are varieties of cereal Barley with two rows of spikelets, and with six rows; the former is the kind generally grown in England. Barley is less liable to disease than Wheat; but it also is subject to the burnt-ear, smut, and ergot. It was used for bread almost universally in ancient times. Ruth gleaned by the maidens of Boaz "unto the end of the Barley harvest, and of Wheat harvest;" and the "six measures" which Boaz put into her veil were of Barley. In the time of the famine of Samaria Elisha is commissioned to promise the miraculous plenty of "a measure of fine flour for a shekel, and two measures of Barley for a shekel." Barley bread is still the regular food of vast numbers of the continental poor; but with us its chief use is for malt. Fowls thrive best when fed with Barley; and Pearl Barley, a preparation of the grain with the skin and part of the seed ground off, is valuable for making barley-water and for puddings. The Barley-grasses are plants of no account.

The Wall Barley (H. murinum, *fig.* 13), is common in waste places; its rough leaves, fringed glumes, and bristly spike readily distinguish it.

The Meadow Barley has shorter awns, but in other respects closely resembles the Wall Barley; and the Sea Barley is only distinguished by its glaucous foliage.

The Oat family (Avéna), are characterised by their loose panicle and large flowers, and by the strong twisted awn with which one of their paleæ is armed.

There is a foreign species called the Animal Oat, the parts of which are so curiously affected by the atmosphere, that when the spikelets are placed on the hand or on the ground they curl, and twist, and leap about as if they were endowed with life.

The Oat-grasses we have examined, so only the cereal Oat remains to be discussed (A. satíva). This grain is the easiest of cultivation, it will grow upon any kind of land; the best samples are produced in Scotland and Friesland. Oatmeal is used by the poor to a great extent in Scotland, Ireland, and the north of England, both for making porridge and for oatcakes. In Switzerland they bake the oatmeal, and then boil it in broth. Oats are exceedingly valuable as food for horses.

The Rye family (Secále), have naked seeds on a flat ear furnished with awns, something like Barley. The straw is solid, being filled with pith, which makes it valuable for thatching as well as for litter. It grows on poor lands which are unfit for Wheat. In former times it was much cultivated in England and called "Meslin," from a word meaning *mixed*, because this grain is the best when mingled with other grains. Excellent bread is made of one-third of Rye-meal to two-thirds of Wheat, and a small quantity mixed with coarse flour makes the best brown bread. It is very much cultivated in Holland, where it forms as important an ingredient in gin as our malt does in beer. In England it is mostly sown as a green crop, and fed-off early in the spring with sheep. The land will bear a crop of Turnips or Potatoes afterwards. It is subject to the same diseases as Wheat, and the ergot affects it to a much greater extent. In a period of famine in France the starving poor in one district were driven to eat the ergotted Rye. All who ate it became miserably ill, their limbs rotting off before death. At first this was supposed to be the effect of witchcraft; but, experiments having been made with ergotted Rye upon animals, it was at length satisfactorily proved that the fungus-poisoned corn was the cause of the malady. It swells the grain to twice its natural length, causing it to assume the form of a black horn.

The Millet cereals, both those belonging to the Soft-grasses and to the Panick-grasses, we have already considered, so that the only important foreign cereals which remain to be described are the Rice and Sugar Cane.

The Rice family has two glumes to each flower, two nearly equal paleæ, six stamens, and two styles.

The common Rice (Orýza satíva), has a diffuse panicle; it is indigenous about lakes in India. A Rice field is much more prolific than a corn field; for one acre will produce from thirty to sixty bushels. Rice is extensively cultivated in India, China, and in Spain, as well as in North and South Carolina, and Georgia. The best ground for Rice fields is low land by the side of rivers, for they require to be well watered, and in this case they need no manure. The uses of Rice are manifold; it is the principal food of great numbers in India, and is largely used in Europe as an article of diet in soups, curries, puddings, &c. Starch is made from Rice as well as from Wheat. The straw of Rice as well as that of Wheat is used in the manufacture of bonnets.

The Sugar Canes (Sacchárum), are very handsome plants; their height varies from eight to twenty feet, and the stem is divided at short intervals by angular joints, from each of which long narrow leaves sprout. The spikelets are composed of two membranous glumes, and florets with two awnless transparent paleæ; there are three stamens and two styles. All the florets are feathery and fertile.

The Indian Sugar Cane has been introduced into the South of Europe, the Canary Islands, and the West Indies; it produces a fine sweet Sugar.

The Chinese Sugar Cane is a larger species, and valuable, because the Canes are hard enough to resist the attacks of the white ants; this Cane abounds in sap.

The natives of Bengal make pens of the small Brown Cane, and they use its stems for making screens and such light articles.

In India Canes are employed for thatching.

The Sugar Cane is propagated by cuttings. Holes are made in rows three feet apart, and two feet between each pair of holes; two or more cuttings are laid at the bottom of each hole, and covered with two inches depth of earth. In about a fortnight the sprouts begin to appear. The cuttings are planted about March in the West Indies, and the harvest time varies from August to November. Several crops may be raised from the same roots. Old planters used to renew the Canes once in three years, but it need not be done so often. The Canes are carried from the plantations as they are cut and thrown into a mill; there they are crushed, and the juice expressed, and received into troughs. It runs next into a pan called a clarifier, where it is kept at a certain heat by a fire placed underneath, but not allowed to boil. During this process it is tempered with lime, and skimmed, and then it is drawn off as a clear yellow liquid; it is then exposed to the air in open vessels, and poured from one to another. It is left to percolate slowly through the spongy stem of a Water Plantain: thus it forms into crystals. After this it is ready to be packed in casks for exportation.

The time of gathering in the fruits of the field is, and ought to be, a time of thankfulness and rejoicing. But, alas! the teeming harvest of the Rice field is too often collected amidst the cries of the oppressed slaves. Longfellow associates slavery with this harvest in his poem, "The Slave's Dream."

> "Beside the ungathered Rice he lay,
>   His sickle in his hand;
> His breast was bare, his matted hair
>   Was buried in the sand.
> Again, in the mist and shadow of sleep,
>   He saw his native land."

But this is not universally the case. Many a rich harvest of Rice is gathered in by free men, and we hope many a hearty thanksgiving arises to the Lord of all, who giveth meat in due season. Seed-time and harvest are full of deep meaning to the thoughtful mind; their constant recurrence is a proof of the faithfulness of God, who promised that "seed-time and harvest, summer and winter, should not fail." Many are the touching pictures of harvest which are presented to us in Scripture. Miss Howitt enumerates some of them in her poem of "Corn Fields."

> "I feel the day, I see the field,
>   The quivering of the leaves;
> And good old Jacob and his house
>   Binding the yellow sheaves;
> And at the very hour I seem
> To be with Joseph in his dream.
>
> "I see the fields of Bethlehem,
>   And reapers many a one,
> Bending with their sickle stroke,
>   And Boaz looking on;
> And Ruth, the Moabitess fair,
> Among the gleaners stooping there.

"Again, I see a little child,
   His mother's sole delight,
God's living gift of love unto
   The kind good Shunamite;
To mortal pangs I see him yield,
And the lad bears him from the field.

"The sun-bathed quiet of the hills,
   The fields of Galilee,
That eighteen hundred years agone
   Were full of corn, I see;
And the dear Saviour takes his way,
'Mid ripe ears on the Sabbath-day.

"O golden fields of bending corn,
   How beautiful they seem!
The reaper-folk, the piled-up sheaves,
   To me are like a dream;
The sunshine and the very air
Seem of old time, and take me there."

Fields of ripe corn are ever the signals of joy and plenty. The Psalmist sings, "The valleys are covered with corn, they shout for joy." Seed-time is typical of the period of faith and patience, " Blessed are they that sow beside all waters." "They that sow in tears shall reap in joy." "Behold, the husbandman waiteth for the precious seed of the earth, and hath long patience for it, until he receives the early and the latter rain." And in the same way the harvest is type of the crowning time of faith—"Be ye also patient, stablish your hearts, for the coming of the Lord draweth nigh." "There shall be a handful of corn in the earth high upon the mountains, the fruit shall shake like Lebanon, and they of the city shall flourish like Grass of the earth."

But the most solemn association of the corn field is with the resurrection:—"That which thou sowest is not quickened except it die; and when thou sowest thou sowest not that body which shall be, but bare grain, it may chance of Wheat or of some other grain, and God giveth it a body as it pleases Him, and to every seed his own body." "Lift up your eyes and look on the fields, for they are white already to harvest." "And he that reapeth receiveth wages, and gathereth fruit unto life eternal, that both he that soweth and he that reapeth may rejoice together." "The harvest is the end of the world, and the reapers are the angels."

The Grasses conclude the families of Flowering Plants, and with them our little collection is complete. You have each of you expressed to me the great pleasure and profit which you have derived from your rambles in search of wild flowers; and all around you testify to the increased health of body and mind which has followed this pursuit. That these benefits may continue and increase I would entreat you to persevere in your research, endeavouring to obtain the many plants yet wanting in our collection, examining more carefully the structure and habits of those you already know, and entering upon the new field of FLOWERLESS PLANTS. Although Fanny is about to take up her abode in London, she will have country trips and holidays, in which she may study and collect *Ferns;* while Edward, in his many woodland wanderings, may do the same service for Mosses and Lichens. Seaweed and Fungi shall be my care; and if we are spared to meet again in another year or two, I hope our collection of Flowerless Plants, with their uses, charms, and associations, will equal in bulk and interest the present result of our "Rambles in Search of Wild Flowers."

INDEX.

A.	PAGE
Acer	68
ACERACEÆ	68
Aceras	277
Achillea	178
Aconite, Winter	19
Aconitum	19
ACORACEÆ	300
Acorus	300
Acotyledons	5
Adonis	14
Adoxa	143
Ægopodium	140
Æthusa	140
Agrimony	161
Agrimony, Hemp	169
Agrostemma	54
Agrostis	315
Aira	317
Ajuga	224
Alchemilla	102
Alder	262
Alexanders	138
Alisma	297
ALISMACEÆ	296
Alkanet	201
Allseed	119
Allium	290
Almond group	93
Alnus	262
Alopecurus	314
Althæa	65
Alyssum	32
AMARYLLIDACEÆ	281
Amaryllis	281
AMENTACEÆ	258
Ammophila	327
AMYGDALEÆ	93
Anacharis	273
Anagallis	235
Anchusa	201
Andromeda	187
Anemone	12
Angelica	137
Anthemis	178
Anthers	6
Anthoxanthum	313
Anthriscus	130
Anthyllis	87
Antirrhinum	211
Apargia	160
Apium	138

	PAGE
APOCYNACEÆ	194
APOCYNEÆ	194
Apple group	105
Aquilegia	19
ARACEÆ	299
ARALIACEÆ	142
Arbutus	188
Archangel	225
Arctium	165
Arenaria	57
Aristolochia	249
ARISTOLOCHIACEÆ	249
Arrow-grass	297
Arrow-head	297
Artichoke, Jerusalem	159
Artemisia	170
Arum	299
Arundo	327
ASARABACEÆ	249
Asarum	249
Ash	190
Ash, Mountain	106
Asparagus	286
Aspen	260
Asperula	153
Asphodel, Bog	294
Asphodel, Scottish	292
Aster	172
Astragalus	87
Athamanta	136
Atriplex	242
Atropa	208
Avena	325
Avens	96
Awlwort	31
Azalea	187

B.	
Ballota	224
Balm, wild	228
Balsam tribe	73
BALSAMINACEÆ	73
Baneberry	19
Barbarea	35
Barberry tribe	27
Barley	332
Barrenwort	27
Bartsia	215
Basil, wild	228
Bastard Toad Flax	248

	PAGE		PAGE
Beaked Parsley	130	Bupleurum	140
Beak Rush	306	Bur Dock	165
Bean tribe	80	Bur Marigold	169
Bearberry	188	Bur Reed	298
Beard-grass	315	Burnet	102
Bedstraw	151	Burnet, Salad	102
Beech	262	Burnet Saxifrage	136
Beet	242	BUTOMACEÆ	296
Bellis	175	Butomus	296
Bell-Flower tribe	180	Butter Bur	172
Bennet, Herb	96	Buttercups	14
Bent-grass	316	Butterfly Orchis	275
BERBERIDACEÆ	26	Butterwort	230
Berberis	27	Buxus	251
Beta	242		
Betony	226		
Betula	260	C.	
BETULACEÆ	260		
Bidens	169	Cabbage	38
Bilberry	185	Cakile	31
Bindweed	199	Calamint	223
Birch	260	Callitriche	112
Bird Cherry	94	Calluna	187
Bird's-foot	89	Calyciflorals	78
Bird's-foot Clover	85	Calyx	6
Bird's-foot Lotus	86	Calyx, subclass	7
Bird's-nest tribe	189	Caltha	18
Birthwort tribe	249	Campanula	180
Bistort	244	CAMPANULACEÆ	180
Bitter Cress	33	Campion	52
Bitter Vetch	91	Canary-grass	313
Blackberry	99	Candytuft	32
Black Thorn	94	Canterbury Bell	183
Bladderwort	230	Capsella	29
Blinks, Water	115	CAPRIFOLIACEÆ	148
Bluebottle, Corn	168	Cardamine	33
Bog Asphodel	294	Carduus	166
Bog Orchis	278	Carex	309
Bog Rush	306	Carpinus	264
Borage tribe	202	CARYOPHYLLACEÆ	50
BORAGINACEÆ	200	CARYOPHYLLEÆ	50
BORAGINEÆ	202	Carraway	140
Box	251	Carlina	167
Bramble	99	Carline Thistle	167
Brassica	38	Carpel	6
Briar, Sweet	103	Carrot, wild	131
Broccoli	39	Carum	140
Brome-grass	324	Castanea	263
Bromus	324	Catchfly	51
Brooklime	217	Catkin tribe	257
Brookweed	237	Cat Mint	227
Broom	82	Cat's-ear	161
Broom, Butcher's	286	Cat's-tail-grass	314
Broom Rape tribe	210	Celandine, Greater	25
Bryony, Black	285	Celandine, Lesser	16
Bryony, White	115	CELASTRACEÆ	78
Buckbean	198	Celery, wild	138
Buckwheat	245	Centaurea	168
Buckthorn tribe	79	Centaury	196
Bug'e	224	Centunculus	237
Bugloss	203	Ceratophyllum	113
Bugloss, Viper's	200	Cerastium	58
Bullace	94	Chærophyllum	130
Bulrush	307	Chaffweed	237
Bunium	132	Chamomile	178

INDEX.

	PAGE		PAGE
Charlock	39	Cornish Moneywort	216
Cheiranthus	37	Corolliflorals	145
Chelidonium	25	Corolla, subclass	146
CHENOPODIACEÆ	240	Corolla, forms of	145
Chenopodium	240	CORYLACEÆ	264
Cherry, wild	93	Corylus	264
Chervil, wild	130	Cotton-grass	308
Chestnut, Sweet	263	Cotton Plant	63
Chestnut, Horse	69	Cotton Thistle	167
Chickweed tribe	55	Cotyledon	121
Chickweed, common	56	Couch-grass	331
Chickweed, Mouse-ear	58	Cowbane	137
Chickweed, Umbelliferous	55	Cowberry	185
Chickweed Winter-green	237	Cow Parsnip	141
Chickweed, Water	278	Cowslip	232
Chicory	164	Cow Tree	193
Chlora	197	Cow Wheat	213
Chrysanthemum	177	Crab Apple	105
Chrysocoma	169	Crambe	31
Chrysosplenium	124	Cranberry	184
Cicely, Sweet	132	Cranesbill	69
Cichorium	164	CRASSULACEÆ	120
Cicuta	137	Cratægus	107
Cineraria	174	Crepis	162
Cinquefoil	97	Cress, Bitter	33
Cinquefoil, Marsh	98	Cress, Rock	34
Circæa	111	Cress, Winter	35
CISTACEÆ	44	Cress, Water	35
Cistus	44	Cresworts	29
Cistus, Gum	44	Crithmum	134
CISTINEÆ	43	Crocus	280
Clary	222	Crosswort	151
Cleavers	152	Crowberry tribe	249
Clematis	12	Crowfoot tribe	14
Clinopodium	228	CRUCIFERÆ	28
Cloudberry	100	Cuckoo-Flower	33
Clove Pink	50	Cuckoo Pint	299
Clover	84	CUCURBITACEÆ	114
Club Rush	307	Cudweed	171
Cnicus	166	Currant tribe	123
Coidium	140	Cuscuta	200
Cochlearia	30	Cyclamen	234
Cockle, Corn	54	Cynodon	316
Cock's-foot-grass	322	Cynoglossum	205
Codlins and Cream	110	Cynosurus	322
Colchicum	291	CYPERACEÆ	306
Colocynth	115	Cyperus	306
Coltsfoot	172	Cyphel	58
Columbine	19	Cytisus	82
Comfrey	202		
COMPOSITÆ	158		
Conium	137	D.	
Convallaria	286		
CONVOLVULACEÆ	199	Dactylis	322
Convolvulus	199	Daffodil	281
CONIFERÆ	266	Dahlia	159
Conyza	175	Daisy, common	175
Coralwort	33	Daisy, Michaelmas	172
Cord-grass	322	Daisy, Ox-eye	177
CORNACEÆ	143	Dame's Violet	39
CORNEÆ	143	Dandelion	163
Cornel tribe	143	Daphne	248
Corn Marigold	177	Darnel	326
Corn Cockle	54	Datura	209
Corn Salad	155	Daucus	131

INDEX.

	PAGE
Deadly Nightshade	208
Dead Nettle	225
Delphinium	20
Devil's-bit	157
Dewberry	101
Dianthus	50
Dicotyledons	5
Digitaria	316
Digitalis	211
DIOSCOREACEÆ	285
DIPSACACEÆ	155
Dipsacus	155
Dock	246
Dodder	200
Dogstander	174
Dog's-tail-grass	322
Dog's-tooth-grass	316
Dog-wood	143
Doronicum	175
Dove's-foot	70
Draba	31
Dropwort	95
Dropwort, Water	133
DROSERACEÆ	47
Drosera	47
Dryas	96
Duckweed tribe	301
Dutch Myrtle	262
Dyer's Green-weed	81
Dyer's Rocket	44
Dyer's Weed	43

E.

	PAGE
Earthnut	132
Echium	200
Eglantine	103
ELATINACEÆ	49
ELATINEÆ	49
Elder	147
ELÆAGNEÆ	247
Elecampane	175
Eleocharis	307
Elm tribe	256
EMPETRACEÆ	249
Empetrum	249
Enchanter's Nightshade	111
Endive	158
Endogens	271
Epilobium	109
Epimedium	27
Epipactis	278
Erica	185
Erigeron	172
Eriophorum	308
Erodium	72
Ervum	89
Eryngium	131
Erysimum	37
Erythræa	197
Euonymus	78
EUPHORBIACEÆ	250
Euphorbia	250
Eupatorium	169

	PAGE
Euphrasia	216
Eve's Cushion	125
Evening Primrose	111
Everlasting Pea	90
Everlasting, Pearly	171
Eyebright	215
Exacum	197
Exogens	270

F.

	PAGE
Fagus	262
Fedia	155
Fennel	138
Fescue-grass	322
Festuca	322
Feverfew	177
Field Madder	153
Figwort	213
Finger-grass	316
Fir Rape	189
Fir tribe	266
Flax seed	67
Flax tribe	66
Fleabane	172
Fleawort	175
Flixweed	36
Flowering Rush	296
Fool's Parsley	140
Forget-me-not	203
Forms of corolla	145
Forms of fruits	116
Forms of Inflorescence	127
Forms of leaves	40
Forms of stem	271
Foxglove	211
Foxtail-grass	314
Fragaria	99
Frankenia	49
FRANKENIACEÆ	49
Fraxinella	72
Fraxinus	190
Frogbit	272
Fruit, forms of	116
Fuller's Teasel	155
Fumaria	26
FUMARIACEÆ	25
Fumeworts	25
Fumitory	26
Furze	81

G.

	PAGE
Galanthus	283
Galeopsis	225
Galeobdolon	225
Gale, Sweet	262
Galingale	306
Galium	151
Garlic	290
Genista	81
GENTIANACEÆ	194
GENTIANEÆ	195

INDEX.

	PAGE		PAGE
Gentian tribe	195	Grass, Tassel	303
Gentian, Lesser	196	Wheat	331
Gentianella	197	Wrack	304
GERANIACEÆ	69	Green-weed	81
Germander	224	Gromwell	201
Germen	6	GROSSULARIACEÆ	123
Geum	96	Grossularia	123
Gillyflower	50	Ground Ivy	227
Gipseywort	221	Groundsel	173
Glasswort	243	Gueldres Rose	149
Glaucium	25		
Glaux	237		
Glechoma	227	H.	
Globe-Flower	17		
GLUMACEÆ	304	Hair-grass	317
Glumals	7	Hard-grass	328
Gnaphalium	171	Harebell	181
Goat's-beard	160	Hare's-ear	140
Golden Moss	122	Hare's-tail-grass	326
Golden Rod	173	Hardhead	168
Golden Saxifrage	124	Hartwort	141
Goldilocks	169	Hawkbit	160
Good-King-Henry	240	Hawksbeard	162
Goodyera	277	Hawkweed	162
Gooseberry tribe	123	Hawthorn	107
Goosefoot tribe	240	Hay, French	89
Goose-grass	152	Hay-green	174
Gorse	81	Hazel	262
Gourd tribe	114	Heath-grass	321
Goutweed	140	Heath-pea	91
GRAMINACEÆ	312	Heath, Sea	49
Grass, Barley	332	Heath tribe	185
Beard	315	Hedera	143
Bent	315	Hedge Mustard	36
Brome	324	Hedge Parsley	132
Canary	313	Hedysarum	89
Cat's-tail	314	Hellebore	18
Cock's-foot	322	Helleborine	278
Cord	322	Hemlock	137
Couch	331	Hemlock, Water	137
Darnel	326	Hemp	254
Dog's-tail	322	Hemp Agrimony	169
Dog's-tooth	316	Hemp Nettle	225
Fescue	322	Henbane	208
Finger	316	Henna	114
Foxtail	314	Hepatica	19
Hair	317	Heracleum	141
Hard	328	Herb Bennet	96
Hare's-tail	326	Herb Christopher	19
Heath	321	Herb Paris	285
Holy	319	Herb Robert	70
Lyme	328	Herminium	277
Mat	313	Herniaria	119
Melic	319	Hesperis	39
Meadow	319	Hibiscus	63
Millet	315	Hieracium	162
Moor	319	Hierochloe	319
Oat	325	HIPPOCASTANACEÆ	60
Panick	317	Hippocrepis	89
of Parnassus	48	HIPPURIDACEÆ	112
Quaking	321	Hippuris	112
Rye	326	Holcus	318
Soft	318	Hollyhock	63
Sweet	320	Holly tribe	189
Sweet Vernal	313	Holly, Sea	131

	PAGE		PAGE
Holy-grass	319	Knappia	315
Honesty	32	Knapweed	168
Honewort	137	Knawel	119
Honeysuckle	148	Knot-grass family	118
Honeywort	151	Knot-grass, Whorled	119
Hop	253	Knot-grass	244
Hordeum	332		
Horehound, Black	224		
Horehound, White	225	**L.**	
Hornbeam	264		
Horned Pondweed	304	LABIATÆ	220
Horned Poppy	25	Labiate tribe	220
Hornwort	113	Laburnum	83
Horse-chestnut	69	Lactuca	161
Horseradish	30	Lady's Finger	87
Horse-shoe Vetch	89	Lady's Mantle	102
Hottonia	234	Lady's Tresses	278
Houseleek	121	Lagurus	326
Humulus	253	Lamb's Lettuce	155
Hutchinsia	29	Lamium	225
Hyacinth, Starch	289	Lapsana	164
Hyacinth, wild	288	Larkspur	20
HYDROCHARIDACEÆ	272	Lathræa	210
Hydrocharis	272	Lathyrus	90
Hydrocotyle	140	Laurel	248
HYPERICACEÆ	61	Laurel, Spurge	248
Hypericum	61	Lavender	220
Hypochœris	161	Leaves, forms of	40
Hyoscyamus	208	LEGUMINOSÆ	80
		LEMNACEÆ	301
		Lemna	301
I.		LENTIBULARIACEÆ	230
		Leontodon	164
Ilex	189	Leonurus	225
ILICINEÆ	189	Leopard's-bane	175
ILLECEBRACEÆ	118	Lepidium	29
Illecebrum	119	Lesser Gentian	196
Impatiens	73	Lettuce, Lamb's	155
Imperatoria	141	Lettuce, Wall	161
Inflorescence	127	Lettuce, wild	162
Insect Orchis	277	Leucojum	284
Inula	175	Ligusticum	141
IRIDACEÆ	279	Ligustrum	190
Iris	279	LILIACEÆ	285
Isatis	32	Lily, Lent	281
Isnardia	111	Lily tribe	285
Ivy	142	Lily, Water	22
Ivy, Ground	227	Lily of the Valley	286
Ixia	280	Lime Tree	60
		Limosella	204
		LINACEÆ	66
J.		Lintbell	66
Jack-go-to-bed-at-noon	159	Linum	66
Jacob's Ladder	198	Ling	187
Jasione	183	Linnæa	150
Jujube	79	Listera	277
JUNCACEÆ	292	Lithospermum	201
JUNCAGINACEÆ	297	Littorella	239
Juncus	292	Lobelia	184
Juniper	268	LOBELIACEÆ	184
		Lolium	326
		London Pride	126
K.		London Rocket	36
		Lonicera	148
Kale, Sea	31	Loosestrife	236

	PAGE		PAGE
Loosestrife, Purple	113	Milkwort	48
LORANTHEÆ	148	Milium	315
Lords and Ladies	299	Millet-grass	315
Lotus	87	Mint	222
Lovage	141	Mint, Cat	227
Lousewort	214	Mistletoe	146
Lucerne	84	Mœnchia	55
Lunaria	32	Moneywort	236
Lungwort	201	Moneywort, Cornish	216
Luzula	294	Monkshood	20
Lycopus	221	Monocotyledons	5, 270
Lyme-grass	328	Monotropa	189
Lysimachia	236	Montia	115
LYTHRACEÆ	113	Moor-grass	319
Lythrum	113	Moschatel	143
		Mother o' Millions	211
		Motherwort	225
M.		Mountain Ash	106
		Mouse-ear Chickweed	58
Madder	151	Mousetail	17
Madder, Field	153	Mudwort	204
Malaxis	278	Mugwort	170
MALVACEÆ	63	Mullein	219
Mallow	64	Mustard, Wild	39
Mallow, Marsh	65	Muscari	289
Mallow, Tree	65	Myosotis	203
Man Orchis	277	Myrica	262
Maple	68	Myriophyllum	112
Mare's-tail	112	Myrrhis	132
Marigold, Corn	177	Myrtle, Dutch	262
Marigold, Marsh	18		
Marjoram	223		
Marrubium	225	N.	
Marsh Cinquefoil	98		
Marsh Mallow	65	NAIADACEÆ	302
Marsh Marigold	18	Narcissus	281
Masterwort	141	Nardus	313
Mat-grass	313	Narthecium	294
Mathiola	38	Nasturtium	35
Matricaria	178	Navelwort	121
May	107	Nectary	7
Meadow-grass	319	Neottia	278
Meadow Rue	12	Nepeta	227
Meadow Saffron	291	Nettle, Dead	225
Meadow Sweet	95	Nettle, Hemp	225
Meconopsis	25	Nettle tribe	252
Medick	85	Nightshade, Enchanter's	111
Medlar	107	Nightshade, Deadly	208
Melampyrum	213	Nightshade, Woody	207
MELANTHACEÆ	291	Nipplewort	164
Melica	319	Nuphar	23
Melic-grass	319	NYMPHACEÆ	22
Melittis	228		
Melilot	84		
Mentha	222	O.	
Menyanthes	198		
Menziesia	187		
Mercury	251	Oak	265
Mespilus	107	Oat	333
Meum	138	Oat-grass	325
Mezereon	248	ŒNOTHERACEÆ	109
Mignonette, wild	44	Œnanthe	133
Milfoil, Water	112	OLEACEÆ	247
Milk Parsley	141	Oleaster tribe	247
Milk Vetch	88	Olive tribe	190

INDEX.

	PAGE
Ononis	83
Onopordon	167
One-lobed class	5, 270
Ophrys	277
Orache	242
ORCHIDACEÆ	274
Orchis, Bog	278
Orchis, Insect	277
Orchis, Man	277
Orchis, Musk	277
Ornithogalum	289
Ornithopus	89
Origanum	223
OROBANCHACEÆ	210
Orobanche	210
Orobus	91
Orpine	122
OXALIDACEÆ	73
Oxalis	73
Ox-eye Daisy	176
Oxlip	233
Ox-tongue	160
Oxyria	247
Oxytropis	87

P.

	PAGE
Panick-grass	317
Panicum	317
Pansy	46
PAPAVERACEÆ	23
PAPILIONACEÆ	80
Park leaves	62
Paris, Herb	285
Parnassia	48
Parietaria	253
PARONYCHIEÆ	118
Parsley, Beaked	130
Parsley, Fool's	140
Parsley, Hedge	132
Parsley, Milk	141
Parsley, Stone	136
Parsley, wild	138
Parsnip, Cow	141
Parsnip, Water	136
Parsnip, wild	141
Parsons in the Pulpit	299
Pasque-Flower	13
Pastinaca	141
Pea, Everlasting	90
Pea, Sweet	90
Pea tribe	80
Pear, wild	105
Pearlwort	54
Pellitory	253
Penny Cress	29
Penny Royal	222
Pœony	21
Peplis	113
Pepper Saxifrage	140
Pepperwort	29
Periwinkle	194
Persicaria	244
Petallids	7, 274

	PAGE
Petalless subclass	240
Petals	6
Petasites	172
Peucedanum	141
Phalaris	313
Pheasant's-eye	14
Phleum	314
Phyteuma	183
Pickpocket	56
Picris	160
Pignut	132
Pimpernel	235
Pine tree	266
Pinguicula	230
Pink tribe	50
Pinus	266
PISTIACEÆ	301
Pistils	6
Pisum	91
Plantain	238
Plantain, Water	296
Plantago	238
Plantaginea	238
PLANTAGINACEÆ	238
Ploughman's Spikenard	173
Plum, wild	94
PLUMBAGINEÆ	238
Poa	320
POLEMONIACEÆ	198
Pollen	6
Polygala	48
POLYGALACEÆ	48
POLYGONACEÆ	244
Polygonum	244
Polypogon	315
Pondweed	302
Pondweed, Horned	304
Poplar	259
Poppy, Horned	25
Poppy, Welsh	25
Poppy tribe	23
Populus	259
PORTULACACEÆ	115
Potamogeton	302
Potentilla	96
Poterium	102
Prenanthes	161
Primrose	231
Primula	231
PRIMULACEÆ	231
Privet	190
Prunella	228
Prunus	93
Pulmonaria	201
Purple Loosestrife	113
Purslane tribe	115
Purslane, Water	113
Pyrola	189
Pyrus	105

Q.

	PAGE
Quaking-grass	321
Quercus	265

INDEX.

R.

	PAGE
Radiola	67
Radish	39
Ragged Robin	53
Ragwort	173
Rampion	180
Rampion, Roundheaded	183
RANUNCULACEÆ	11
Ranunculus	11
Rape	38
Raphanus	39
Raspberry	100
Rattle, Yellow	214
Receptacle subclass	7
Reed	327
Reed-mace tribe	298
RESEDACEÆ	43
Rest Harrow	83
RHAMNACEÆ	79
Rhinanthus	214
Rhodiola	123
Rhyncospora	306
Ribes	123
Rib-grass	238
Rice	334
Rocket, London	35
Rocket tribe	44
Rock Cress	34
Rock Rose	44
Rod, Golden	173
Roots, forms of	74
ROSACEÆ	103
Rose, Gueldres	149
Rose, Rock	44
Rose of Sharon	45
Rose tribe	103
Rose-root	123
Rottböllia	328
Rowan tree	107
Rubia	151
RUBIACEÆ	150
Rubus	99
Rue, Garden	72
Rue, Meadow	12
Rumex	245
Ruppia	303
Rupturewort	119
Ruscus	286
Rush, Beak	306
Rush, Bog	3, 6
Rush, Bull	307
Rush, Club	307
Rush, Flowering	296
Rush, Spiked	307
Rye	333

S.

	PAGE
Saffron	281
Saffron, Meadow	291
Sage, wild	222
Sagina	55
Sagittaria	297

	PAGE
Saintfoin	89
Salad, Burnet	102
SALICACEÆ	257
Salicornia	243
Sallow	257
Salsola	243
Salep	275
Salix	257
Saltwort	243
Salsafy	160
Salvia	221
Samolus	237
Samphire	134
Sambucus	147
Sandalwood tribe	248
Sandwort	57
Sanguisorba	102
Sanicle	131
SANTALACEÆ	248
Saponaria	52
Sauce-alone	37
Saw-wort	165
SAXIFRAGACEÆ	124
Saxifrage	124
Saxifrage, Pepper	140
Saxifrage tribe	124
Scabiosa	157
Scabious, Sheep's	183
Scandix	130
Scilla	287
Scirpus	306
Scleranthus	119
Scorpion-grass	203
Scotch Fir	266
Scottish Asphodel	292
SCROPHULARIACEÆ	204, 211
SCROPHULARIÆ	213
Scurvy-grass	30
Scutellaria	228
Sea Buckthorn	247
Sea Cabbage	38
Sea Heath	49
Sea Holly	131
Sea-kale	31
Sea Pea	91
Sea Plantain	238
Sea Lavender	238
Sea Milkwort	237
Sea Reed	327
Sea Rocket	31
Sea stock	38
Sedge	309
Sedge, Sweet	300
Sedum	120
Self-Heal	228
Selinum	141
Sempervivum	121
Senebiera	32
Senecio	173
Serratula	165
Service Tree	106
Sesleria	319
Shamrock	85
Sheep's-bit	183
Shepherd's Needle	130

	PAGE		PAGE
Shepherd's Weather-glass	235	Subularia	31
Shepherd's Purse	29	Sugar Cane	335
Sherardia	154	Sulphurwort	141
Shoreweed	239	Sundew	47
Sibbaldia	98	Sweet Briar	103
Sibthorpia	216	Sweet Grass	320
Silene	51	Sweet Vernal-grass	313
Silver Weed	96	Sweet Sedge	300
Sinapis	39	Sycamore	68
Sison	137	Symphytum	202
Sisymbrium	36		
Sium	136		
Skullcap	228	**T.**	
Sloe	94		
Smyrnium	138	TAMARICACEÆ	49
Snakeweed	244	Tamarisk	48
Snapdragon	211	Tamus	285
Sneezewort	178	Tanghinea Nut	193
Snowdrop	283	Tanacetum	169
Snowflake	284	Tansy	169
SOLANACEÆ	208	Tare	89
Solanum	207	Tarragon	159
Soapwort	52	Tassel-grass	303
Soft-grass	318	TAXACEÆ	267
Solidago	173	Taxus	267
Solomon's Seal	287	Teasel	155
Sorrel	246	Teesdalia	29
Sorrel, Mountain	247	Teucrium	224
Sorrel, Wood	73	Thalamiflorals	7
Southernwood	170	Thalictrum	12
Sow-bread	235	Thesium	248
Sow Thistle	161	Thistle, Carline	167
Sparganium	298	Thistle, Cotton	167
Spartina	322	Thistle, Scotch	167
Spearwort	16	Thlaspi	29
Speedwell	216	Thorn, White	167
Spergula	55	Thorn, Black	94
Spignel	140	Thorn Apple	209
Spikenard, Ploughman's	175	Thrift	237
Spike Rush	307	Thrincia	160
Spindle Tree	78	Thyme	223
Spiræa	95	THYMELACEÆ	248
Spurge tribe	250	Thyme, Water	273
Spurge Laurel	248	TILIACEÆ	60
Spurrey	55	Tilia	121
Squill	287	Toad Flax	212
Squinancywort	153	Toad Flax, Bastard	248
Statice	237	Tofieldia	292
Stachys	226	Toothwort	210
Starwort	112	Tordylium	141
St. John's Wort tribe	61	Torilis	132
St. Peter's Wort	62	Tormentil	98
Stamens	6	Touch-me-Not	73
Star of Bethlehem	289	Tower Mustard	35
Stellaria	56	Tragopogon	159
Stigma	6	Traveller's Joy	12
Stitchwort	56	Treacle Mustard	37
Stock	38	Tree Mallow	65
Stonecrop	120	Trefoil	84
Stone Parsley	136	Trefoil, Bird's-foot	87
Storksbill	72	Trientalis	237
Strapwort	118	Triglochin	297
Stratiotes	272	TRILLIACEÆ	285
Strawberry, wild	99	Triodia	321
Strawberry Tree	188	Trifolium	84

	PAGE		PAGE
Triticum	330	Water Awlwort	31
Trollius	17	Water Blinks	115
TROPÆOLACEÆ	72	Water Chickweed	278
True-love	285	Water Cress	35
Tulip	291	Water Dropwort	133
Turnip	39	Water Flag	279
Turritis	35	Water Hemlock	137
Tussilago	172	Water Lily	22
Tutsan	62	Water Milfoil	112
Twayblade	277	Water Parsnip	136
Two-lobed class	5	Water Plantain	296
TYPHACEÆ	298	Water Purslane	115
		Water Soldier	272
		Water Starwort	112
		Water Thyme	273
U.		Water Violet	234
		Waterwort tribe	49
Ulex	81	Weather-glass, Shepherd's	235
ULMACEÆ	256	Weasel-snout	225
Ulmus	256	Weld	43
UMBELLIFERÆ	129	Wheat	330
Umbelliferous Chickweed	55	Wheat-grass	331
Urtica	252	Whin	81
URTICACEÆ	252	White Beam	106
Utricularia	230	White Bryony	115
		White Horehound	225
V.		White Rot	140
		Whitlow Grass	81
VACCINIEÆ	185	Whortleberry	185
VALERIANACEÆ	154	Wild Balm	228
VALERIANEÆ	154	Wild Basil	228
Valerian tribe	154	Wild Chamomile	178
Valisneria	272	Wild Parsley	138
Vanilla	274	Wild Rocket	35
Venation of leaves	40, 271	Willow	257
Venus's Looking-glass	191	Willow-herb	109
Verbena tribe	229	Wind-Flower	13
VERBENACEÆ	229	Winter Cress	35
Vernal-grass	313	Winter Chickweed	237
Verbascum	219	Winter-green	188
Veronica	216	Woad	32
Vervain	229	Woodbine	148
Vetch	88	Woodruff	153
Vetch, Bitter	91	Wood Rush	294
Vetch, Milk	87	Wood Sorrel	73
Vetchling	90	Wormwood	170
Viburnum	149	Woundwort	226
Vicia	88		
Villarsia	198		
Vinca	194	Y.	
VIOLACEÆ	45		
Violet	45	Yam tribe	285
Violet, Dame's	39	Yarrow	178
Violet, Water	234	Yellow Archangel	225
Viper's Bugloss	200	Yellow Pimpernel	236
VISCACEÆ	146	Yellow Rattle	214
Viscum	146	Yellowwort	197
		Yew	267
W.			
		Z.	
Wallflower	37		
Wart Cress	32	Zannichellia	304
Wartwort	250	Zostera	304

LONDON:
PRINTED AT THE HORTICULTURAL PRESS, 171, FLEET STREET.

First-class Weekly Illustrated Gardening Publication,
Price Threepence; Stamped, Fourpence.

THE JOURNAL OF HORTICULTURE,
COTTAGE GARDENER,
AND
COUNTRY GENTLEMAN,
EDITED BY GEORGE W. JOHNSON, F.R.H.S.,
AND
ROBERT HOGG, LL.D., F.L.S.,

Assisted by a Staff of the Best Writers on Practical Gardening, and Numerous Correspondents engaged in the pursuit of Horticulture and Rural Affairs.

THIS Publication some time since commenced a New Series of that old-established and popular Periodical "THE COTTAGE GARDENER," permanently increased to Thirty-two Pages, and richly illustrated with Wood Engravings in the highest style of the Art.

The subjects treated on embrace every department of Gardening, and Rural and Domestic Economy; from the small plot and allotment of the Cottager to the villa garden of the Country Gentleman and Merchant, the grounds of the Parsonage-house, and the more extensive establishments of the Nobility and Gentry.

The Horticultural Department consists of all Out-door and In-door Operations of the Fruit, Flower, and Kitchen Garden; Notices of all the New Fruits, Flowers, and Vegetables; Arboriculture, and more particularly Fruit Tree Culture and Pomology; Landscape Gardening and Garden Architecture; Descriptions of all the Newest Inventions in Garden Structures, Tools, and Implements; and a Detail of Work to be done in each Department during every week in the year.

In Rural and Domestic Economy it treats of the Farm and Poultry-yard; Allotment Farming; the Dairy; the Pigeon-house; and Rabbit and Bee-keeping. The Treatment of Soils, Manures, Cropping, and Rotation of Crops. Brewing; Wine-making; Vegetable Cookery, and the Preserving of Fruits and Vegetables.

Natural History and Botany, so far as they relate to Gardening and Husbandry, are amply treated on; and embrace Zoology, Geology, Mineralogy, Meteorology, and Physiological, Structural, Systematic, and Popular Botany.

Biographies and Portraits of the most celebrated Horticulturists.

Reviews of New Books relating to the above subjects Reports of Horticultural and Poultry Societies' Meetings throughout the country; and Scientific Notices.

To ADVERTISERS the JOURNAL OF HORTICULTURE will be found a valuable and effective medium, from the extent of its circulation among the middle and higher classes.

A SPECIMEN NUMBER FREE BY POST FOR FOUR STAMPS.

JOURNAL OF HORTICULTURE & COTTAGE GARDENER OFFICE,
171, FLEET STREET, LONDON, E.C.;
And to be had of all Booksellers and at the Railway Stalls.

Monthly, Price 1s.; with Four Coloured Illustrations, THE

WILD FLOWERS OF GREAT BRITAIN,

Botanically and Popularly Described, with Copious Notices of their History and Uses,

By ROBERT HOGG, LL.D., F.L.S.,

And GEORGE W. JOHNSON, F.R.H.S.,

EDITORS OF THE "JOURNAL OF HORTICULTURE AND COTTAGE GARDENER."

Illustrated by Coloured Drawings of all the Species,

BY CHARLOTTE GOWER.

Vol. I., with Eighty Coloured Plates, is now ready, elegantly bound in cloth, gilt extra, price 21s.

Beautiful Coloured Engravings of the New Flowers and Fruits appear in

THE FLORIST AND POMOLOGIST,

A PICTORIAL MONTHLY MAGAZINE OF

FLOWERS, FRUITS, AND GENERAL HORTICULTURE,

CONDUCTED BY

ROBERT HOGG, LL.D, F.L.S.,

ASSISTED BY

Mr. THOMAS MOORE, F.L.S.,

AND NUMEROUS ABLE CONTRIBUTORS.

ONE SHILLING MONTHLY.

The Volumes for 1862 and 1863, with Twenty-three Highly Coloured Illustrations, price 14s., bound in cloth, gilt extra, are now ready.

MANUALS FOR THE MANY.

GARDENING FOR THE MANY. Threepence.
ALLOTMENT FARMING FOR THE MANY. Threepence.
BEE-KEEPING FOR THE MANY. Fourpence.
GREENHOUSES FOR THE MANY. Sixpence.
KITCHEN GARDENING FOR THE MANY. Fourpence.
FLOWER GARDENING FOR THE MANY. Fourpence.
FRUIT GARDENING FOR THE MANY. Fourpence.
FLORISTS' FLOWERS FOR THE MANY. Fourpence.
POULTRY BOOK FOR THE MANY. Sixpence.
WINDOW GARDENING FOR THE MANY. Ninepence.
MUCK FOR THE MANY. Threepence.
RABBIT BOOK FOR THE MANY. Sixpence.
HEATING MANUAL. Sixpence.

⁎ Any of the above can be had post free for an additional postage stamp.

LONDON: "JOURNAL OF HORTICULTURE AND COTTAGE GARDENER" OFFICE, 171, FLEET STREET, E.C.

And to be had of all Booksellers and at the Railway Stalls.

www.ingramcontent.com/pod-product-compliance
Lightning Source LLC
Chambersburg PA
CBHW020228240426
43672CB00006B/455